iGlobal
Educational Services
Believe.Inspire.Transform.

Calculus
Study Guide

ISBN-13: 978-1-944346-62-1

To order, contact:

iGlobal Educational Services
PO Box 94224
Phoenix, AZ 85070
Website: www.iglobaleducation.com
Fax: 512-233-5389

HOW TO USE THIS STUDY GUIDE

iGlobal Educational Services created this study guide to help you review mathematical concepts that may help you increase your knowledge of Calculus topics.

This study guide should be used to supplement strong and viable curriculum that encourages differentiation for all diverse learners. They can be used at home, in tutoring sessions, or at school.

TABLE OF CONTENTS

CALCULUS: FUNCTIONS AND THEIR APPLICATIONS

Content Description

In this session, we will discuss quite a number of functions together with their forms after transformations. We will wind up with the application of all these functions discussed.

MATH TOPICS

- Calculus 104.1 Linear and quadratic equations (Reference 1.11).
- Calculus 104.2 Applications (Reference 1.12).
- Calculus 104.3 Functions including special functions, such as rational functions, exponential, square roots, logarithmic (Reference 1.13).
- Calculus 104.4 Transformation of parent functions (Reference 1.14).
- Calculus 104.5 Applications (Reference 1.15).

INTRODUCTION

It is a common practice that if the price of a packet of pieces of candies is $2.3, then for each packet that is added, we add $2.3, to get the total cost. This can be done for whichever number of packets that are required. When a person has a given budget, then formula can be derived to calculate if the budget is enough to buy the required amount of packets or not. Likewise, when we want to design a bridge with curved arches, we first come up with a formula to guide us. These among other daily activities requires formulas. These may be linear quadratic, exponential, logarithmic among others. Thus, it is worthy familiarizing with these formulas.

SECTION 104.1: LINEAR AND QUADRATIC EQUATIONS

A linear function is an equations that is in the form $ax + by + c = 0$. However, the most common representation is the slope-intercept form which is $y = f(x) = mx + c$ where m is the slope and c the $y-$ intercept of the function.

The slope is the rise divided by the run of the graph while the $y-$ intercept is the $y-$ value where the graph intercepts the $y-$ axis.

The graph of a linear function is a straight line. When the value of $c = 0$, then the graph passes through the origin.

When m is negative, the graph slopes to the right and when it is positive, it slopes to the left.

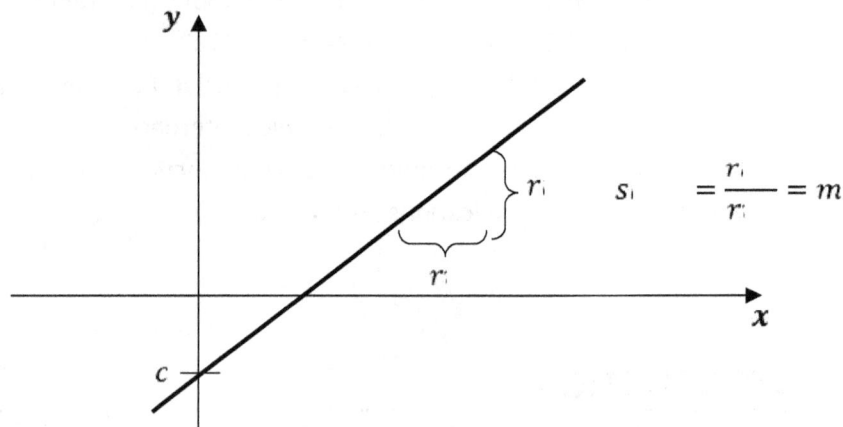

Finding linear equation

We can find the equation of the linear function given the coordinates of at least two points on the line or given a point and the slope.

Example 1
Given that a line passes through $(0, 2)$ and $(5, 1)$, determine its equation.

Solution
The slope is
$$\frac{rise}{run} = \frac{1 - 2}{5 - 0} = -\frac{1}{5}$$

Let (x, y) be on the line too. Using the definition for slope, we have
$$\frac{y - 2}{x - 0} = -\frac{1}{5}; \qquad 5(y - 2) = -x; \quad 5y - 10 = -x$$

Thus, $5y = 10 - x$ or $y = 2 - \dfrac{x}{5}$

The function is

$$f(x) = 2 - \dfrac{x}{5}$$

Example 2

Determine a function $f(x)$ describing all points on a line whose slope is 2.5 given that $(-1, -7)$ is one of the points.

<u>Solution</u>

The slope is

$$\dfrac{rise}{run} = 2.5 = \dfrac{5}{2}$$

If (x, y) is also on the line as $(-1, -7)$, we have

$$\dfrac{y+7}{x+1} = \dfrac{5}{2} \quad ; \quad 2y + 14 = 5x + 5 \quad ; \quad 2y = 5x - 9 \quad ; \quad y = \dfrac{5}{2}x - \dfrac{9}{2}$$

Thus, the function is

$$f(x) = \dfrac{5}{2}x - \dfrac{9}{2} \quad or \quad f(x) = 2.5x - 4.5$$

Quadratic function

This is a function which is given by $f(x) = ax^2 + bx + c$ where, a, b, c are constants with a being non zero. A quadratic function has the highest powers on its variable as 2, hence, it has two roots, that is, the values of x when $f(x) = 0$.

The functions has a graph in form of a parabola. The graph has a vertex and is symmetric about a line that passes through the vertex.

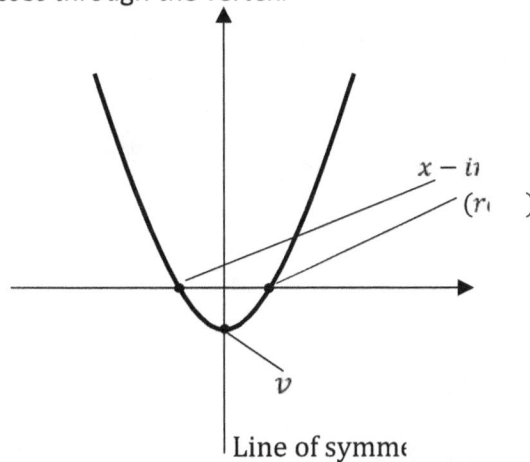

$x - i$
$(r_. \quad)$

v

Line of symme

The vertex form of a quadratic function is given by $f(x) = a(x - h)^2 + k$ which is derived from the standard form $f(x) = ax^2 + bx + c$ using completing square method.

Roots quadratic function

These are values of the independent variable that reduces the function to zero. The quadratic function $f(x) = ax^2 + bx + c$ has a root at $x = r$ if $f(r) = ar^2 + br + c = 0$.

The roots are determined using factor method, completing square method, graphical method, substitution method and quadratic formula among others. The latter is the best one.

The quadratic formula for determining the roots of $f(x) = ax^2 + bx + c$ is

$$x = \frac{-b \pm \sqrt{b^2 - 4ac}}{2a}$$

Example 1

Find the values of x such that $h(x) = 0$ if $h(x) = 2 - 3x^2 + 9x.$

Solution

In standard form, the function $h(x) = 2 - 3x^2 + 9x$ is $h(x) = -3x^2 + 9x + 2$. Thus,

$a = -3, b = 9$ and $c = 2.$

Using the quadratic formula, we have

$$x = \frac{-b \pm \sqrt{b^2 - 4ac}}{2a} = \frac{-9 \pm \sqrt{81 + 14}}{-6} = \frac{-9 \pm \sqrt{95}}{-6} = \frac{-9 \pm 9.747}{-6}$$

$x = -0.1245, \qquad x = 3.125$

Example 2

Identify the vertex of the function $f(x) = 3x^2 - 12x + 5$

Solution

We change the function into vertex form using completing square

$$f(x) = 3x^2 - 12x + 5 = (3x^2 - 12x) + 5 = 3(x^2 - 4x) + 5$$

We complete the square in the bracket to get

$$f(x) = 3(x^2 - 4x) + 5 = 3(x^2 - 4x + 4 - 4) + 5 = 3(x^2 - 4x + 4) + 3(-4) + 5$$
$$f(x) = 3(x^2 - 4x + 4) - 7 = 3(x - 2)^2 - 7$$

Hence, the vertex form is $f(x) = 3(x - 2)^2 - 7$

Example 3

Identify the line of symmetric of the function $x = f(y) = -2y^2 + 8y + 7$

Solution

We first convert the function to vertex form

$$x = f(y) = -2y^2 + 8y + 7 = (-2y^2 + 8y) + 7 = -2(y^2 - 4y) + 7$$

Completing the square in the bracket, we get

$$-2(y^2 - 4y) + 7 = -2(y^2 - 4y + 4 - 4) + 7$$
$$f(x) = -2(y^2 - 4y + 4) + 8 + 7$$
$$= -2(y - 2)^2 + 15$$

The vertex is $(y, x) = (2, 15)$

Since the square is on y , the graph is horizontal, hence the line of symmetry is horizontal too. The line of symmetry is $y = 2$.

Section 104.2: Applications

Having looked at the details of linear and quadratic functions, we can now discuss a few applications of the concept in real life.

Revenue and cost function

Revenue from the sale of a commodity is a linear function $R(x) = px$ where p the price of a commodity is and x is a number of items.

Cost function is a function that determines the cost of a commodity that has both fixed and variable function. For instance, water has fixed cost and variable cost. If the fixed cost of water every month is f and the variable cost is px where p is the cost per unit gallon and x, is the numberof gallons, then the cost function is

$$C(x) = px + f$$

The marginal cost is the slope of the cost function. Thus, it is the value of p while the average cost is the $\dfrac{C(x)}{x}$.

Example 1
The cost of a pizza is $28.
(i). Write a function that shows the cost of any number of pizza.
(ii). Use the function to determine the cost of 3 pizzas.
(iii). If Jenkins has a budget of 100 to buy pizza, write a function that will determine the maximum number of pizzas that he can buy.

Solution

(i). Let the cost of pizza be p. x Pizzas will cost px.
Thus, the total cost of pizza is $R(x) = px$
Since $p = 28$, the function is $R(x) = 28x$

(ii). The number of pizzas is 3, thus, $x = 3$
   The total cost of these three is $R(3) = 28(3) = \$84$

(iii). The cost of pizza is $R(x) = 28x$

Since he has $100, he can buy a maximum of x_t where $28x_t = 100$, x_t being an integer number.
The function is $f(x_t) = 28x_t - 100$, x_t being an integer number

Example 2

The cost function of electricity in a month is given by the function $C(x) = 32x + 1.4$ where x is in kilowatt hours.

(i). Find the marginal cost of the electricity.
(ii). What would be the cost of electricity of Pauline used 1.25 kilowatt hours of electricity in a month?
(iii). Find paulines average cost of electricity.

Solution

(i).The marginal cost is the slope of the cost function. Comparing $C(x) = 3.2x + 1.4$, with $y = mx + c$, the slope is $m = 3.2$ kilowatt hours.

(ii). If the electricity consumed is 1.25 megawatt hours, we substitute this for x
$$C(1.25) = 3.2(1.25) + 1.4 = 5.4$$

The cost is $5.4

(iii) Cost of electricity is $C(1.25) = 5.4$

Average cost is $\dfrac{C(x)}{x} = \dfrac{5.4}{1.25} = \$4.32$

Focusing

Linear functions, also referred as linear model are used in focusing future values of an occurrence. For instance, if the relationship with the number of forest coverage (in square miles) of a state is a function time (in years), then we may have a linear model representing this and can be used to estimate the possible forest cover in future.

Example 3

The forest cover of a state is each year is determined by the function $F(t) = 52t + 349$.
(i). What is the expected forest cover in 2030 if the forest cover at 2013 was 349.
(ii). What is the expected increase in forest cover from 2017 to 2025?

Solution

The forest cover is given by the function $F(t) = 52t + 349$

(i). From 2013 to 2030, we have a difference of $2030 - 2013 = 17$ years
The forest cover in 2030 is expected to be $F(17) = 52(17) + 349 = 1233$ sq. miles

(ii). From 2013 to 2017 and 2025 is 4 years and 12 years respectively.

The increase in forest cover is
$$F_{2025}(t) - F_{2017}(t) = (52(12) + 349) - (52(4) + 349)$$
$$= 973 - 557 = 416$$

The increase is forest cover is 416 sq. miles

Bridges

Quadratic curves are used in designing arches for buildings, bridges and similar structural materials. If the constrains of the height and width of the arch is provided, then a quadratic function can be achieved. This quadratic function will help in coming up with the actual design of the arch.

Example 4

The figure below shows a front view of a bride that id made up of parabolic sections. For the designers to come up with a good design of the view, they have to come up with equations representing each line. However, the equation of the first parabola is required then the rest will be determined without any easily. Determine the equation describing the first parabola taking the corner at the lowest left end as the origin.

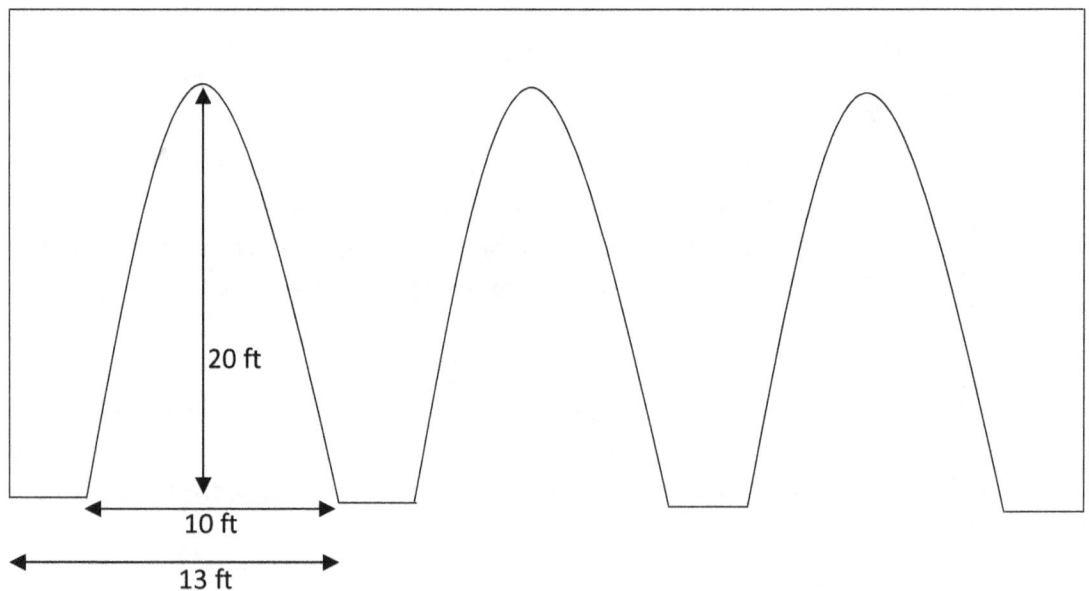

Solution

We can come up with three points on the graph as shown below. The points are A, B and C. The coordinates of these points are

$A(3,0)$, $B(13,0)$ and $C(8,20)$.

R Remember, C the the vertex and it is at the middle of the end points of the parabola, in terms of $x - \mathbf{axis, thus, 3 + 5 = 8}\ (x - \mathbf{coordinate})$

The quadratic function is given by $y = ax^2 + bx + c, c \neq 0$

We carry out substitution for each point.

At $A(3,0)$,

$0 = 9a + 3b + c$ ....(i)

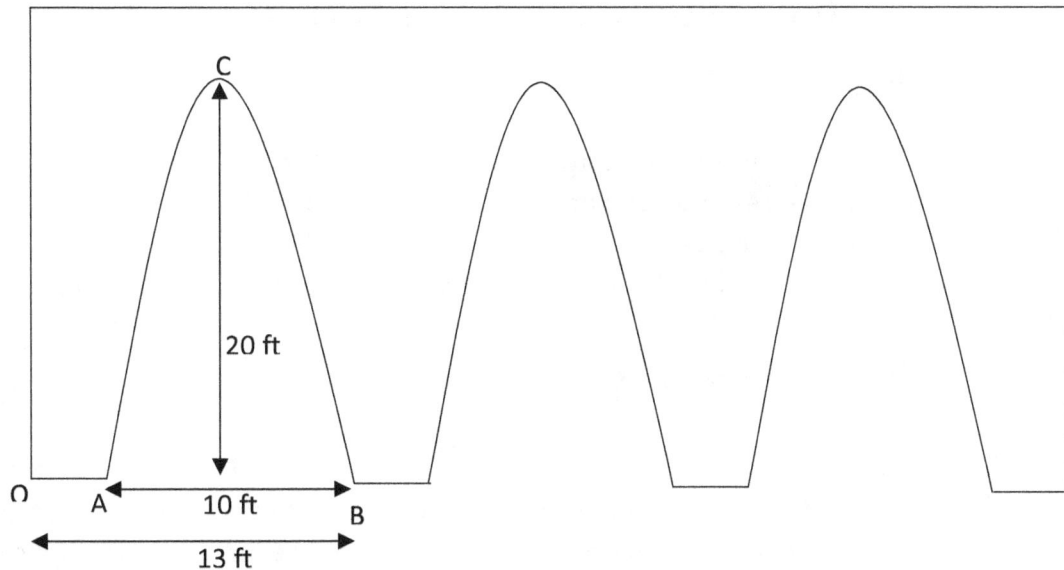

At $B(13,0)$

$0 = 169a + 13b + c = 0$ ....(ii)

At $C(8,20)$

$20 = 64a + 8b + c$ ....(iii)

Subtracting (i) from (ii), we get $0 = 160a + 10b; \quad 16a + b = 0; \quad b = -16a$

Subtracting (i) from (iii), we get $20 = 55a + 5b$

Substituting for b , we get $20 = 55a + 5(-16a) = 55a - 80a = -25a;$

This implies that $20 = -25a, \qquad a = -0.8, b = 12.8$

From (i), $c = -9a - 3b = -9(-0.8) - 3(12.8) = -31.2$

Hence, the parabola is given by
$f(x) = -0.8x^2 + 12.8x - 31.2$

Section 104.3 Functions including special functions, such as rational functions, exponential, square roots, logarithmic

Functions are relations between two sets called the domain (the set of all inputs) and the range (the set of all outputs) where one value of a domain is not assigned more than one value in the range. We would like to consider the most special types of functions, that is, rational functions, exponential functions, square root functions, and logarithmic functions.

Rational functions

These are functions in the form $g(x) = \dfrac{f(x)}{h(x)}$ where $h(x)$ can never be zero and $f(x)$ and $h(x)$ are polynomials. The most common properties that describe these functions is the intercepts and the asymptotes. The intercepts are points on the main axes where the graph passes while the asymptotes are straight lines where the graph approaches but does not intersect nor touch.

$x -$ Intercepts

This is a point where the function intersects the $x -$ axis. At this point, the coordinate of y is zero.

Thus, at this point $g(x) = \dfrac{f(x)}{h(x)} = 0$, implying that $f(x) = 0$.

$y -$ intercept

This is a point where the function intersects the $y -$ axis. At this point, the coordinate of x is zero. This is also called the root of a function.

Thus, at this point $g(0) = \dfrac{f(0)}{h(0)}$, implying that $x = 0.$

Asymptotes

$g(x) = \dfrac{f(x)}{h(x)}$ has a vertical asymptote at $x = a$ where $h(a) = 0.$

For a horizontal and oblique, asymptote, we have three cases

(i). If the degree of $f(x)$ is less than that of $h(x)$, the function has a horizontal asymptote at $y = 0$.

(ii). If the degree of $f(x)$ is equal to that of $h(x)$, the function has a horizontal asymptote at $y = \dfrac{h}{k}$, where h and k are the leading coefficients of $f(x)$ and $h(x)$ respectively.

(ii). If the degree of $f(x)$ is more than that of $h(x)$, the function has no horizontal asymptote but has an oblique asymptote $y = ax + b$, **where** $ax + b$ is the quotient when $f(x)$ is divided by $h(x)$.

Example 1

Find all features of

$$f(x) = \frac{16x^2 - 1}{x^2 - 4}$$

<u>Solution</u>

The $x-$ intercept

It occurs when $f(x) = 0$. Thus

$$\frac{16x^2 - 1}{x^2 - 4} = 0; \; 16x^2 - 1 = (4x)^2 - 1^2 = 0$$

$$(4x)^2 - 1^2 = (4x - 1)(4x + 1) = 0$$

$4x - 1 = 0$ implies $4x = 1; \; x = \dfrac{1}{4} = 0.25$

$4x + 1 = 0$ implies $4x = -1; \; x = -\dfrac{1}{4} = -0.25$

The $x-$ intercepts are $x = 0.25$ and $x = -0.25$

The $y-$ intercept

It occurs when $x = 0$.

$$f(0) = \frac{16(0)^2 - 1}{(0)^2 - 4} = \frac{1}{4} = 0.25$$

$y-$ intercept is 0.25

Asymptotes

For vertical asymptote, we have $x^2 - 4 = 0$.

$x^2 - 4 = x^2 - 2^2 = (x-2)(x+2) = 0; \quad x - 2 = 0 \text{ or } x + 2 = 0$

Thus, $x = 2$ and $x = -2$ are the vertical asymptotes

Horizontal asymptote: The denominator and the numerator have the same degree, hence, the horizontal asymptotes is the quotient of the leading diagonals. The horizontal asymptote is $y = 16$.

Exponential function

These are function of the form $f(x) = ab^{kx}$ where a, b and k are non-zero contents with $b \neq 1.$ When $k > 0$, the function is decreasing and when $k < 0$, the function decreasing. The function has $y = 0$ as the horizontal asymptote.

The functions are defined on the whole of the real axis, however, their range is the positive values only.

The most common exponentials function is $f(x) = ae^{kx}$, in this case, $b = e$ where e is an irrational number called Euler's number. $e = 2.718281 \ldots$

The functions is used to describe natural occurrences. In this case, k is referred to as the growth rate when $k > 0$ or as decay rate when $k < 0$ and a , the initial value.

Exponential laws

$a^x \times a^y = a^{x+y}$

$a^x \div a^y = a^{x-y}$

$a^0 = 1, \qquad a^1 = a; \ = a^{-y} = \dfrac{1}{a^y}$

$(a^x)^y = a^{xy}$

$a^x = a^y$ implies that $x = y$

Example 2

Solve the following

$$5^x + 5^x = 50$$

Solution

If we let 5^x be y , then we have $2y = 50$; $y = 25$

Hence $5^x = 25 = 5^2$

This implies that $x = 2$ since exponential function is a one to one function

Example 3

Solve for t is $\dfrac{2}{32^x} = 128^{-2x+1}$

Solution

We express 32 and 128 in base 2.

$32 = 2 \times 2 \times 2 \times 2 \times 2 = 2^5$

$128 = 32 \times 2 \times 2 = 2^5 \times 2^2 = 2^{5+2} = 2^7$

Upon substitution, we have

$\dfrac{2^1}{2^{5x}} = 2^{-7(2x+1)}$; $\quad 2^{1-5x} = 2^{-7(2x+1)}$ *(division law)*

Since exponential function is one to one, we have and $a^x = a^y$ implies that $x = y$, we have

$$1 - 5x = -7(2x + 1) = -14x - 7$$

Thus, we have

$$9x = -8; \quad x = -\frac{8}{9}$$

Square roots function

These are functions where the variable is under a square root function. That is $f(x) = \sqrt{x}$.

The square root functions are only defined on positive values including zero only. However, their range is the whole of the real number line.

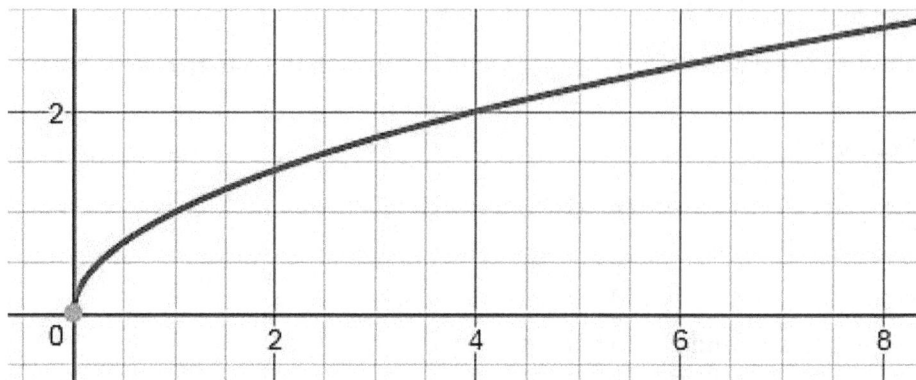

Example 4

Find the zero of the following $f(x) = \sqrt{2x + 1}$

Solution

The zero of a function is a number of x where $f(x) = 0$.

$f(x) = \sqrt{2x + 1} = 0;$ implying that $2x + 1 = 0$; $2x = -1$; $x = -0.5$

Logarithmic functions

These are functions of the form,

$f(x) = \log_a x$

where a is a non-negative constant not equal to 0 nor 1.

Logarithmic functions and exponential functions are inverses of each other hence, they are a reflection of one another at $y = x$.

Given the

$f(x) = \log_a x$

Its equivalent exponential notation is $x = a^{f(x)}$

Or if $\log_a x = b,$ **then** $x = a^b$

Logarithmic functions are not defined for non-negative values of x nor 0 but its range is the whole of the real line. It has a vertical asymptote at $x = 0$.

When $a = 10$, we call the functions, a common logarithm function, while when it is e , we call it a natural logarithm function and it is abbreviated $\ln x = \log_e x$.

To work with logarithmic functions, we use the laws of logarithms.

Laws of logarithms

$\log_a a = 1; \ \log_a 1 = 0,$

$\log_a ab = \log_a a + \log_a b$

$\log_a \dfrac{a}{b} = \log_a a - \log_a b$

$x\log_a b = \log_a b^x$

If $\log_a d = \log_a b$, then $d = b$

Example 5

Solve for x

(i). $\log_{10} 8 - \log_{10}(x + 3) = \log_{10} 2x + 12$

(ii). $\log_x 3 - \log_x 15 = 2$

Solution

(i).Using multiplication law, we have $\log_{10} 8 + \log_{10}(x + 3) = \log_{10} 2x + 12$ being equivalent to $\log_{10} 8(x + 3) = \log_{10} 2x + 12$

Since logarithm function is injective, we have

$8(x + 3) = 2x + 12$

$8x + 24 = 2x + 12$

$6x = -12$

$x = -2$

Remember, this answer is valid since $\log_{10}(x + 3)$ and $\log_{10} 2x + 12$ are defined when $x = -2$

(ii). The base of the logarithm is x , hence $1 = \log_x x, 2 = 2\log_x x$

Upon substitution, we have

$\log_x 3 - \log_x 15 = 2\log_x x$

By the property $x\log_a b = \log_a b^x$, we have

$$\log_x 3 - \log_x 15 = \log_x x^2$$

Using division law of logarithms, we get

$$\log_x 3 - \log_x 15 = \log_x \frac{3}{15} = \log_x 0.2 = \log_x x^2$$

By infectivity of logarithm functions, we get $0.2 = x^2$

Thus, $x = \pm\sqrt{0.2} = \pm 0.4472$

Since we cannot have a negative base, we have $x = 0.4472$

Section 104.4: Transformation of parent functions

Parent function is a function is a very basic functional representation of any given function. Most transformations can be described in term of a parent function. The examples of these parent functions are $y = x, y = x^2, y = e^x, y = \sqrt{x}$ for linear, quadratic, exponential and square root functions respectively.

Algebraic transformations

This is a modification of the parent function. We may have translation, compression, stretch and reflection among others.

Translation (shift)

Translation refers to the slide of a function. Given that the parent functions $y = f(x)$ moves a units to the right , the resultant function is $y = f(x - a)$ and $y = f(x + a)$ if the slide was to the left.

We also have motion in vertical direction.

When the function $y = f(x)$ shifts upwards by a units, the resultant is $y = f(x) + a$ and $y = f(x) - a$ if it shifted downwards by the same value.

Example 1

Translate the following function 3 units to the left and 4 units downwards. $y = 2x^2 + 4x + 1$

<u>Solution</u>

A translation of 3 units to the left implies $y = f(x + 3)$

Thus, the function becomes $y = f(x + 3) = 2(x + 3)^2 + 4(x + 3) + 1$

$$y = 2(x^2 + 6x + 9) + 4x + 12 + 1$$

$$y = 2x^2 + 12x + 18 + 4x + 12 + 1 = 2x^2 + 16x + 31$$

A translation of 4 units downwards implies $y = f(x) - 4$

Thus, the function becomes $f(x) - 4 = 2x^2 + 16x + 31 - 4$

Thus, we have $f(x) = 2x^2 + 16x + 27$

Reflection

We consider only two mirror lines. Reflection about $y-$ axis transforms $y = f(x)$ to $y = f(-x)$ while that of $x-$ axis transforms the same function to $y = -f(x)$

Stretch

Again, our interest here is vertical and horizontals stretch. A function $y = f(x)$ is stretch vertically by a scale factor of $a, a > 1$, to get $y = af(x)$ horizontally by a scale factor of b , to get $y = f(bx)$ where $0 < b < 1$.

Example 2

Identify the transformation on the parent function of $y = 2\sqrt{4x + 4}$

<u>Solution</u>

We express the function into known transformation forms. The parent function is $y = \sqrt{x}$.
$y = 2\sqrt{4x + 4} = 2\sqrt{0.3(x + 1)}$

We have a horizontal translation of 1 unit to the left to get $y = \sqrt{x + 1}$

Then a horizontal stretch of scale factor 0.3 to get $y = \sqrt{0.3(x + 1)}$

Finally, a vertical stretch on scale factor 2 units to get $y = 2\sqrt{0.3(x + 1)}$

Example 3

Reflect the following function about the y axis; $f(x) = e^{2x-3} + 1$

Solution

A reflection about $y - axis$ leads to a function $y = f(-x)$.

Thus, $f(x) = e^{2x-3} + 1$, becomes,

$f(-x) = e^{2(-x)-3} + 1 = f(x) = e^{-2x-3} + 1 = \dfrac{1}{e^{2x+3}} + 1$

The function is

$f(x) = e^{-2x-3} + 1 = \dfrac{1}{e^{2x+3}} + 1$

Compression

A function $y = f(x)$ is compressed vertically by a scale factor of $a, \text{for } 0 < a < 1$, to get $y = af(x)$ and horizontally by a scale factor of $b, b > 1$ to get $y = f(bx)$.

Example 4

Perform a reflection of $\text{about } x - axis$, a vertical and horizontal shifts of 5 upwards and 1 unit to the right respectively and a vertical compression of scale factor $\dfrac{1}{3}$ units on $f(x) = x^3$

Solution

A reflection about $x -$ axis gives us $y = -f(x)$; that transforms the function to $y = -x^3$

A vertical shift of 5 units upwards gives us $y = f(x) + 5$; that transforms the function to $y = -x^3 + 5$

A horizontal shift of 1 unit to the right gives us $y = f(x - 1)$; that transforms the function to
$y = -([x-1])^3 + 5 = -(x-1)(x-1)^2 + 5$
$\quad = (-x+1)(x^2 - 2x + 1) + 5 = -x^3 + 2x^2 - x + x^2 - 2x + 1 + 5$
$\quad = -x^3 + 3x^2 - 3x + 6$

A vertical compression by a factor of $\dfrac{1}{3}$ units gives us $y = \dfrac{1}{3}f(x)$

The function becomes

$$f(x) = -\dfrac{1}{3}x^3 + x^2 - x + 2$$

Section 104.5: Applications

Functions have quite a lot of applications in real life. We will list a few of them.

Investment and banking

Exponential and logarithmic functions are applied in determination of future value of invested and banked funds. When a given amount of money is invested or banked where it is supposed to increase with a compound interest rate, then the future value of this amount is a function of t. This function is exponential and can be solved using properties of logarithmic functions.

If P is invested and is supposed to increase at a rate of $r\%$, then the future value of the investment is

$$p\left(1 + \frac{r}{100}\right)^t.$$

Decay and growth process

These are natural processes whose state can be determined using exponential functions and analyzed using logarithmic functions.

The decay of N as a function of time, t is given by $N(t) = N_0 e^{kt}$ where, N_0 is the initial amounts and k the decay rate. When $k < 0$, the process is called a decay process while when $k > 0$, the process is growth process.

Depreciation

This is a reduction of value of an item. Most common items that reduce in value are assets such as motor vehicles, house among others. When the value of these commodities are given, then their future value can be calculated based on the rate of depreciation given.

If the current value is P and they decrease in value at a rate of $r\%$, then the future value after t years is

$$p\left(1 - \frac{r}{100}\right)^t.$$

Work, rate and time

Rational functions are used in solving problems related to work rate and time. A function is determined which is later solved. Its solution are the solutions of the problem at hand.

Design problems

Square root functions are applied in solving design problems involving squares. This is because the length of one side of a square is a square root of its area.

Example 1

A population of a town initially at 2.1 million is growing at a rate of 1.21% every years. Determine the population after 3 years.

Solution

The function is $f(x) = ae^{kx}$ where k is the rate and a, the initial value

Since the population was initially at 2.1 million, we have $a = 2100000$

The rate is $1.21\% = 0.0121$

Since the population is increasing, the rate is positive.

The function becomes
$$f(x) = 2100000e^{0.0121x}$$

After 3 years, we have $x = 3$
$$f(3) = 2100000e^{0.0121 \times 3} = 2177630$$

The population would be 2177630

Example 2

A radioactive substance has a half life of 230 years. If it is initially 30grams, what would be remaining amount after 356 years?

Solution

The process is natural; therefore, we use the function $f(x) = ae^{kx}$

Initial amount is $a = 30$

After 230 years, that when $t = 230 \ years$, the substance would be half the initial one. If the initial is a, the remaining would be $0.5a$.

Upon substitution, we get

$0.5a = ae^{230k}$ or $0.5 = e^{230k}$

Applying natural logarithms, we have $\ln 0.5 = \ln e^{230k}$

Applying the low of logarithms, we have $\ln 0.5 = 230k \ln e$

But $\ln e = \log_e e = 1$

We get $\ln 0.5 = 230k; k = \dfrac{\ln 0.5}{230} = -0.003014$

The function is $f(x) = 30e^{-0.003014x}$

After 356 years, we have

$f(x) = 30e^{-0.003014 \times 356} = 10.26 \; grams$

The remaining substance was 10.26 grams

Example 3

An architect would like to design a square room whose area must $32 \; sq.feet.$ determine a function that would approximate the dimensions of the room, hence use the function to solve the problem.

Solution

Let the dimensions of the room be x by x, then $A(x) = x^2$

To find x, we have $x = \pm\sqrt{A(x)}$ or $x = \sqrt{A(x)}$ for $x > 0$

But since the variable is the area, we have $x(a) = \sqrt{a}$

Thus, for $a = 26$, we have $x(32) = \sqrt{32} = 4\sqrt{2} \; ft$

Thus, the dimensions are $4\sqrt{2}$ ft by $4\sqrt{2}$ ft

Example 4

The price of a motor vehicle is $1200. If it depreciates at a rate of 2.45% every years. What would be its value after 8 years.

Solution

The initial value is $p = \$1200$

Rate is $r = 2.45\% = 0.0245$

$t = 8$ years

The value after 8 years is $\quad A = p\left(1 - \dfrac{r}{100}\right)^t = 1200(1 - 0.0245)^8$

$$= 1200(0.9755)^8 = \$984$$

The value is $\$984$

Section 104 Conclusion

In this lesson, we have discussed various properties of functions; that is linear, quadratic, exponential, rational, and logarithmic and square root functions. We have also went further to discuss their applications in real life situation. We hope that you find it interesting.

CALCULUS: CONTINUITY AND SOME CONICS

Content Description

In this session, we will discuss the concept of continuity as developed by the idea of limits. We will then embark of a study of some of the conic sections.

MATH TOPICS

- Calculus 104.6 Limits (Reference 1.11).
- Calculus 104.6 Continuity (Reference 1.12).
- Calculus 104.8 Parabola (Reference 1.13).
- Calculus 104.9 Hyperbola (Reference 1.14).
- Calculus 104.10 Circle (Reference 1.15).

INTRODUCTION

Continuity is a concept that is used to describe lines that do not have gaps or steps along their channels. This concepts is reinforced by the concepts of limit of functions. Some of the continuous functions include the parabola, the hyperbola, and the circle, which we will discuss, though, with minimal relation to limits and continuity.

Section 104.6: Limits

Consider a function $f(x) = 5x - 9$. We may take values of x closer and closer to 4 but not equal to it. Consider 3.9, 3.99, 3.999, 4.0001, 4.001, 4.01, 4.1

If we come up with a table of values for x and $f(x)$, we have

x	3.9	3.99	3.999	4.0001	4.001	4.01	4.1
$f(x)$	10.5	10.95	10.995	11.005	11.005	11.05	11.5

Taking values closer and closer to a number, say 4 for our case above is termed as taking values approaching the number, say 4. Note that number values are approaching 4 from both sides, the its left hand (on a number line) and its right hand (on the number line).

We now describe what happens to the values of $f(x)$.

When x values approaches 4 from the below or left (denoted $x \to 4^-$), the values of $f(x)$ approaches 11 from the lower side. Also, as x values approaches 4 from above (the right), the values of $f(x)$ approaches 11 from the higher side.

As long as we does not take 4, but we choose a series of values closer and closer to 4, we will be approaching 11. Thus, we say 11 is the limit of the function $f(x) = 5x - 9$ as x approaches 4 and we write

$$\lim_{x \to 4} 5x - 9 = 11$$.

Thus, the limit of a function $f(x)$ as x approaches a number a is a number L such that the values of $f(x)$ approaches L when values of x are taken closer and closer to a. We write
$$\lim_{x \to a} f(x) = L$$

Remember, that we considered both right and left and we were approaching the same number from below and above.

The limit taken from the left hand side, that is, numbers less than 4 gave us values closer and closer but less than 11. This limit is called left-hand side limit and it is denoted

left hand side limit *is* $\lim_{x \to 4^-} 5x - 9 = 11$

In general, the left hand side limit is denoted $\lim_{x \to a^-} f(x)$

Also considering the right hand side, we got values of $f(x)$ closer and closer to but larger than 11. This limit is called the right-hand side limit and it is denoted

right hand side limit is $\lim\limits_{x \to 4^+} 5x - 9 = 11$

In general, the right hand side limit is denoted $\lim\limits_{x \to a^+} f(x)$

The limit of a function is said to exist if we get a number as the limit, otherwise, it does not exists. When the limit exists and is equal to L, then $\lim\limits_{x \to a^+} f(x) = \lim\limits_{x \to a^-} f(x) = L$

If the two, right and left hand side limits are not equal, then the limit does not exists at that particular point.

From the example above, we can determine the limit using a table. We can also do the same using analytic methods (computation). When computations does not work, we use the table or observe the graph.

Limits of functions

Polynomial function
We simplify substitute the value of the independent variable to get the limit.

Example 1
Find the limit of $f(x) = x^3 - 3x^2 + 6x + 1$ as x approaches -1.

Solution.
$\lim\limits_{x \to -1} x^3 - 3x^2 + 6x + 1 = (-1)^3 - 3(-1)^2 + 6(-1) + 1 = -1 - 3 - 6 + 1 = -9$
Thus,

$\lim\limits_{x \to -1} x^3 - 3x^2 + 6x + 1 = -9$

Square root, exponential, logarithmic functions

For these functions, we simply substitute for the independent variable if the value is in their domain if not, the limit does not exists.

Rational functions

We first express the function into its simplest form then carry out substitution. If that does not work, we can also use rationalization using the conjugate of the denominator or the numerator where possible.

Other methods are beyond the scope of the lessons.

Example 2

Find the limit of the following functions

(i). $\lim_{x \to 0} x^n$; n is an integer

(ii). $\lim_{x \to 0} \dfrac{1}{x^n}$; n is an integer

(iii). $\lim_{x \to 0^+} \sqrt{x^3 + 4x - 1}$

(iv) $\lim_{x \to -\frac{1}{2}} \dfrac{2x + 1}{4x^2 - 1}$

Solution

(i). $\lim_{x \to 0} x^n$; The function x^n is a polynomial; hence, we carry out substitution.

$\lim_{x \to 0} x^n = 0^n = 0$ The limit exists since 0 is a number

(ii). $\lim_{x \to 0} \dfrac{1}{x^n}$; n is an integer

When we fix n , say $n = 2$, the vary x , we get the answer.

$\dfrac{1}{x^2}$;

When $x = 1$, the number is 1, when it is 0.5, the number is more than one. It is 2. When it is 0.1, the number is 10. The number keeps on increasing as we move to zero. Thus, Al most at zero, the number will be heading to undefined big "number."

Thus $\lim_{x \to 0} \dfrac{1}{x^n} = \infty$ .The limit does not exists (iii). $\lim_{x \to 0^+} \sqrt{x^3 + 4x - 1}$

We substitute for x

$\lim_{x \to 0^+} \sqrt{x^3 + 4x - 1} = \sqrt{(0)^3 + 4(0) - 1} = \sqrt{-1}$ does not exists

The limit does not exists

(iv) $\lim_{x \to -\frac{1}{2}} \dfrac{2x + 1}{4x^2 - 1}$

We first simplify the function,

$$\frac{2x + 1}{4x^2 - 1} = \frac{2x + 1}{(2x)^2 - 1^2} = \frac{2x + 1}{(2x + 1)(2x - 1)} = \frac{1}{2x - 1}$$

Thus,

$$\lim_{x \to -\frac{1}{2}} \frac{2x + 1}{4x^2 - 1} = \lim_{x \to -\frac{1}{2}} \frac{1}{2x - 1} = \frac{1}{\left(2 \times -\frac{1}{2}\right) - 1} = -\frac{1}{2}$$

Infinite limits

These are limits taken as the value of the independent variable approaches infinity.

We will investigate simply rational and polynomial functions only.

(i). $\lim_{x \to \infty} x^n$; n is an integer

As x becomes biggere and bigger (approaches infinity), x^n becomes bigger even faster, hence it approaches infinity too.

(ii). $\lim_{x \to \infty} \dfrac{1}{x^n}$; n is an integer

As x becomes biggere and bigger (approaches infinity), $\dfrac{1}{x^n}$ becomes smaller and smaller but positive such that it approaches zero.

$$\lim_{x \to \infty} \frac{1}{x^n} = 0$$

Example 3

Evaluate the following limits

(i).

$$\lim_{x \to \infty} \frac{x^2 + x^3 - 4x + 1}{x^3 + 5x - 2}$$

(ii).

$$\lim_{x \to \infty} \frac{x^2 + 2x^3 - 4x + 1}{x^5 + 2x - 7}$$

(iii).

$$\lim_{x \to \infty} \frac{9x^2 + 3x^6 - 4x + 1}{x^5 + 11x - 2}$$

Solution

(i). We divide by the highest power in both the numerator and the denominator.

$$\lim_{x \to \infty} \frac{x^2 + x^3 - 4x + 1}{x^3 + 5x - 2} = \lim_{x \to \infty} \frac{\frac{1}{x} + 1 - \frac{4}{x^2} + \frac{1}{x^3}}{1 + \frac{5}{x^2} - \frac{2}{x^3}}$$

Taking limits, all functions of the form $\frac{1}{x^n}$ approaches zero. Thus, we have

$$\lim_{x \to \infty} \frac{\frac{1}{x} + 1 - \frac{4}{x^2} + \frac{1}{x^3}}{1 + \frac{5}{x^2} - \frac{2}{x^3}} = \frac{1}{1} = 1$$

The limit exists

$$\lim_{x \to \infty} \frac{9x^2 + 3x^6 - 4x + 1}{x^5 + 11x - 2} = 1$$

(ii). We divide by the highest power in both the numerator and the denominator.

$$\lim_{x \to \infty} \frac{x^2 + 2x^3 - 4x + 1}{x^5 + 2x - 7} = \lim_{x \to \infty} \frac{\frac{1}{x^3} + \frac{2}{x^2} - \frac{4}{x^4} + \frac{1}{x^5}}{1 + \frac{2}{x^4} - \frac{7}{x^5}}$$

Taking limits, all functions of the form $\frac{1}{x^n}$ approaches zero. Thus, we have

$$\lim_{x \to \infty} \frac{\frac{1}{x^3} + \frac{2}{x^2} - \frac{4}{x^4} + \frac{1}{x^5}}{1 + \frac{2}{x^4} - \frac{7}{x^5}} = \frac{0}{1} = 0$$

The limit exists

$$\lim_{x \to \infty} \frac{x^2 + 2x^3 - 4x + 1}{x^5 + 2x - 7} = 0$$

(iii). We divide by the highest power in both the numerator and the denominator.

$$\lim_{x \to \infty} \frac{9x^2 + 3x^6 - 4x + 1}{x^5 + 11x - 2} = \lim_{x \to \infty} \frac{\frac{9}{x^4} + 3 - \frac{4}{x^5} + \frac{1}{x^6}}{\frac{1}{x} + \frac{11}{x^5} - \frac{2}{x^6}}$$

Taking limits, all functions of the form $\frac{1}{x^n}$ approaches zero. Thus, we have

$$\lim_{x \to \infty} \frac{\frac{9}{x^4} + 3 - \frac{4}{x^5} + \frac{1}{x^6}}{\frac{1}{x} + \frac{11}{x^5} - \frac{2}{x^6}} = \frac{3}{0} = \infty; \textbf{undefined}$$

Thus, the limit does not exist

Section 104.7: Continuity

This is a term that is used to describe curves or straight lines that are drown without any gap, hole and any form of a step along its channel. Continuity is described by the concepts of limits. A function $y = f(x)$ is said to be continuous at a point say, $x = a$ if

(i). The function is defined at that point. That is $f(a)$ exists

(ii). $\lim_{x \to a} f(x) = f(a)$

A function is continuous on an interval if it is continuous at all points within the interval

A function is continuous on the real number set if it is continuous at all points within the real number set

This definition shows that polynomial functions are continuous for the whole of the real numbers line since their limits are determined by simply substitution.

Let use the above definition to determine the continuity of functions.

Example 1

Determine if the following functions are continuous are the specified points.

(i). $f(x) = x^2 + 6x + 4$ at $x = -3$.

(ii).

$f(x) = \dfrac{3x - 2}{9x^2 - 4}$ at $x = \dfrac{2}{3}$

(iii). $f(x) = \log(5 - x) - \log(2x + 8)$ at $x = 6$

(iv). $f(x) = \begin{cases} -x + 1, & x < 0 \\ 2x + 1, & x \geq 0 \end{cases}$ at $x = 0$

(v). $f(x) = \begin{cases} x, & x < -1 \\ 3, & -1 \leq x < 1 \\ \dfrac{x}{2} + 1, & x \geq 1 \end{cases}$ at $x = -1$

Solution

(i). $f(x) = x^2 + 6x + 4$ at $x = -3$.

$f(-3) = (-3)^2 + 6(-3) + 4 = -5$ the function is defined at $x = -5$

$\lim_{x \to -3} x^2 + 6x + 4 = (-3)^2 + 6(-3) + 4 = -5 = f(-3)$

Thus, the function is continuous at $x = -3$.

(ii).

$$f(x) = \frac{3x - 2}{9x^2 - 4} \quad at \ x = \frac{2}{3}$$

$$f\left(\frac{2}{3}\right) = \frac{3\left(\frac{2}{3}\right) - 2}{9\left(\frac{2}{3}\right)^2 - 4} = \frac{0}{0} = \infty \text{ , this is undefined}$$

As much as the function may have a limit at that point, it is not defined at the same point, hence, not continuous at $x = \frac{2}{3}$.

(iii). $f(x) = \log(5 - x) - \log(2x + 8)$ at $x = 6$

At $x = 6$, we have $f(6) = \log(5 - 6) - \log(2(6) + 8) = \log(-1) - \log 20$

Since $\log -1$ does not exists, the function is not defined at $x = 6$. Thus, the function I not continuous at $x = 6.$

(iv). $f(x) = \begin{cases} -x + 1, & x < 0 \\ 2x + 1, & x \geq 0 \end{cases}$ $\quad at \ x = 0$

The function is defined on different intervals. We use the intervals that are only valid for the number $x = 0$.

Since zero is in the interval, $x \geq 0$, we use the part, $2x + 1$
$$f(0) = 2(0) + 1 = 1$$

Since the function has two parts as x approaches zero from both sides, we consider them all.

Left hand limit
We consider points less than but closer to zero, hence we use the function whose domain is $x < 0$.
$$\lim_{x \to 0^-} -x + 1 = -(0) + 1 = 1$$

Right hand side limit
We consider points more than but closer to zero, hence we use the function whose domain is $x \geq 0$.
$$\lim_{x \to 0^+} 2x + 1 = 2(0) + 1 = 1$$

The two limits are equal; hence, the limit as x approaches 0 exists. $\lim_{x \to 0} f(x) = 1$
Since $\lim_{x \to 0} f(x) = 1 = f(0)$, the function is continuous at $x = 1.$

(v). $f(x) = \begin{cases} 2x, & x < -1 \\ 3, & -1 \leq x < 1 \\ \dfrac{x}{2} + 1, & x \geq 1 \end{cases}$ at $x = -1$

The point, $x = -1$ is in the interval $-1 \leq x < 1$, hence $f(-1) = 3$

Left hand side limit

The left hand side of -1 is in the interval, $x < -1$, thus

$$\lim_{x \to -1^-} f(x) = \lim_{x \to -1^-} 2x = 2(-1) = -2$$

Right hand side limit

The right hand side of -1 is in the interval, $-1 \leq x < 1$, thus

$$\lim_{x \to -1^+} f(x) = \lim_{x \to -1^-} 3 = 3$$

The two sided limits are not equal, $\lim_{x \to -1^+} f(x) = 3 \neq -2 = \lim_{x \to -1^-} f(x)$. Thus the limit $\lim_{x \to -1} f(x)$ does not exists.

This implies that the function is not continuous at $x = -1$.

Section 104.8 Parabola

A parabola is a collection of points is equidistant from both the central point, called the focus, as the line called the directory. The figure below shows the features of a parabola. P is the point on the parabola and line FB is the line of symmetry of the curve. F is the focus and V

The line ABC is the directrix. From the definition, we have that for any point P on the parabola, $FP = PC$.

The equation of the parabola **facing upwards** is given by
$(x - a)^2 = 4p(y - b)$

Where, (a, b) is the vertex, p is the absolute distance from the vertex to the focus and $y = b - p$ the directrix.

Thus, the focus is at $(a, b + p)$

The parabola is facing downwards

The equation is given by
$(x - a)^2 = -4p(y - b)$

Where (a, b) is the vertex and p the absolute distance from the vertex and $y = b + p$ the directrix.

The parabola is facing to the right

When it is facing to the right, the equation is $(y - a)^2 = 4p(x - b)$ where (a, b) is the vertex and p the absolute distance from the vertex to the focus and $x = a - p$ the directrix, (a, b) is the vertex and p the absolute distance from the vertex

The parabola is facing to the left

When facing to the left, the equation is is $(y - a)^2 = -4p(x - b)$ where $x = a + p$ is the directrix, (a, b) is the vertex and p the absolute distance from the vertex.

Example 1

Find the equation of a parabola whose foci is $(2, 7)$ and center is $(2, 5)$.

Solution

The foci is above the vertex hence the parabola is facing upwards. Thus, its equation is given by

$(x - a)^2 = 4p(y - b)$.

The distance between the focus and the vertex is $p = 7 - 5 = 2$

The vertex is $(a, b) = (2, 5)$.

Thus, the parabola is $(x - 2)^2 = 8(y - 5)$

Example 2

Find the equation of the parabola whose foci is $(-2, 0)$ and directrix is $x = 4$.

Solution

The foci is to the left of the directrix hence, the parabola is facing to the left. Its equation is given by $(y - a)^2 = -4p(x - b)$

The distance from the foci to the directrix is $2p = 4 - -2 = 6$; $p = 3$

The vertex is $p = 3$ units to the right of the foci. Thus, the vertex is $(a, b) = (-2 + 3, 0) = (1, 0)$

The equation is $y^2 = -12(x - 1)$

Example 3

Find the directrix of the parabola given by $y^2 = 16 - 8x$

Solution

We express the equation in standard form. That is $y^2 = 8(2 - x) = (-2)4(x - 2)$

The parabola is $(y - 0)^2 = (-2)(4)(x - 2)$

The term with y is squared implying that the parabola is horizontal

The vertex is $(0, 2)$ and the distance from the center to the directrix is $|p| = |-2| = 2$.

The negative on p implies that the parabola is facing in the negative direction of negative $x-$ axis.

Thus, the directrix is $2p$ units which is $4\ units$ to the right of the vertex. Thus, the directrix is $x = 0 + 4$; $x = 4$

Example 4

Find the foci and the vertex of a parabola whose equation is $x^2 = 2y - 4$

<u>Solution</u>

The term with x is squared implying that the parabola is vertical.
Writing it in standard form, we get $x^2 = 2(y - 2)$
The vertex is $(0, 2)$

Thus, $4p = 2; p = \dfrac{1}{2}$

Since p is positive, the parabola is facing upwards, hence the foci is above the vertex.

Thus, the focus half a unit from the vertex.

The coordinates of the foci is $(0, 2 + 0.5) = (0, 2.5)$

Section 104.9: Hyperbola

A hyperbola is a set of points whose positive difference between the distances from the one point to the fixed points called the foci is the same. Each of the two curves that make up a hyperbola is called a Branch. Each Branch has a vertex that lie on one line together with the foci of the hyperbola. The line where the foci and the vertices lie ($F_1 F_2$) is called the transverse axis.

When the hyperbola has the center at (h, k), and P is any point on the parabola, using the properties of the definition of parabola that $PF_2 - PF_1$ is the same for all points, we have the equations

$$\frac{(x - h)^2}{a^2} - \frac{[(y - k)]^2}{b^2} = 1, \text{for a horizontal hyperbola}$$

and

$$\frac{(y - h)^2}{a^2} - \frac{[(x - k)]^2}{b^2} = 1, \text{for a horizontal hyperbola}$$

Where a is the distance from the center to the vertices, b the distance from the vertex to the intersection of the asymptotes and the perpendicular to the vertex. The foci is c units from the center and $c^2 = b^2 + a^2$.

The asymptotes of the hyperbola are $y = \pm \dfrac{a}{b}$, whether it is horizontal or vertical.

When (h, k) = (0, 0) the center of the parabola is at the origin and its equation (horizontal parabola) is

$$\frac{x^2}{b^2} - \frac{y^2}{a^2} = 1$$

.

Example 1

Find the foci of a hyperbola whose equation is $\dfrac{x^2}{3^2} - \dfrac{y^2}{2^2} = 1$

Solution

Since the term having x is positive, we hyperbola is horizontal. Furthermore, from the general equation, $\dfrac{(y-h)^2}{a} - \dfrac{([x-k)]^2}{b^2} = 1$, $h = k = 0$. This implies that the hyperbola is centered at the origin.

From the graph, $a = 3, b = 2$

Since $c^2 = b^2 + c^2 = 3^2 + 2^2;\ \ c = \sqrt{9 + 4} = \sqrt{13} = 3.605$

Thus, the foci are 3.065 units on the left and right of the center.

The foci are $(0, -3.605)$ and $(0, 3.605)$

Example 2

Find the equation of a hyperbola whose center is (2,3) and vertices are at $(2, 5)$ and $(2, 1)$ if the foci is 7 units from the center.

`

Solution

Since the vertices are above and below the center, the hyperbola is vertical.

The distance from the center to the vertex is $a = \sqrt{(2-2)^2 + (5-1)^2} = 4\ units$

The foci is 7 units from the center, $c = 6$. We determine b.

$b = \sqrt{c^2 - a^2} = \sqrt{5^2 - 4^2} = \sqrt{25 - 16} = \sqrt{9}$

$b^2 = 9$

From the general equation,

$$\frac{(y-h)^2}{a^2} - \frac{([x-k)]^2}{b^2} = 1$$

We have the equation of the parabola being ,

$$\frac{(y-3)^2}{4^2} - \frac{([x-2)]^2}{3^2} = 1$$

Example 3

Find the coordinates of the foci and the vertices of the hyperbola given by $\frac{(y+5)^2}{36} - \frac{x^2}{16} = 1$

Solution

We change the equation to its standard form by dividing through by 4.

$$\frac{(y+5)^2}{6^2} - \frac{x^2}{4^2} = 1$$

Thus, the center is $(0,-5)$

Since the term with y is positive, the parabola is vertical. We also have $a = 6$ and $b = 4.$

Thus, $c = \sqrt{a^2 + b^2} = \sqrt{6^2 + 4^2} = \sqrt{52} = 4\sqrt{13} = 14.42$ units

Thus, the foci is 14.42 units above and below the center while the vertices are 6 units below and above the vertex.

The foci are $(0,-5 + 14.42) = (0,9.42)$ and $(0,-5 - 14.42) = (0,-19.42)$

The vertices are $(0,-5 + 6) = (0,1)$ and $(0,-5 - 6) = (0,-11)$

Example 4

Find the coordinates of the center and the vertices of the hyperbola given by $(x-1)^2 - 4y^2 = 4$

Solution

We change the equation to its standard form by dividing through by 4.

$$\frac{(x-1)^2}{4} - \frac{y^2}{1} = 1 \text{ or } \frac{(x-1)^2}{2^2} - \frac{y^2}{1^2} = 1$$

The center is $(1,0)$

The term with x is positive, hence the hyperbola is horizontal.

Comparing the equation with
$$\frac{(y-h)^2}{a^2} - \frac{([x-k)]^2}{b^2} = 1$$

We get that $a = 2$ and $b = 1$.

The vertices are 2 units from the vertex. Since the hyperbola is horizontal, the vertices are two units to the left and right of $(1,0)$.

Thus, the vertices are $(-1, 0)$ and $(3, 0)$.

Section 104.10: Circle

A circle is a collection of points that are equidistant from a fixed point called the center of the circle. The standard equation of the circle is given by
$$Ax^2 + By^2 + Cx + Dy + E = 0$$

Where, A, B, C, D, E are constants with $A = B$.

This equation can be reduced by completing square to center-radius form as $(x-a)^2 + (y-b)^2 = r^2$ where (a, b) is the center and r the radius.

Example 1

Find the equation of a circle passing through $(4, 4)$ and whose center is $(1, 2)$.

Solution

The radius of the circle is the distance from the center to a point on the circle. The distance is determined using Pythagorean Theorem. Thus

$$r = \sqrt{(4-2)^2 + (4-1)^2} = \sqrt{4+9} = \sqrt{13}$$

The equation of the circle is given by $(x-a)^2 + (y-b)^2 = r^2$ where (a, b) is the center and r the radius. Upon substitution, we have

$(a, b) = (1,2)$ and $r = \sqrt{13}$, hence

$$(x - 1)^2 + (y - 2)^2 = \left(\sqrt{13}\right)^2$$

The equation is
$$(x - 1)^2 + (y - 2)^2 = 13$$

Example 2

Find the radius and the center of the following circle given by
$$f(x) = x^2 + y^2 + 6y + x - 1 = 0$$

<u>Solution</u>

We collect the terms having similar variable together then complete the square in each case
$$f(x) = x^2 + y^2 + 6y + x - 1 = x^2 + x + y^2 + 6y - 1 = 0$$
$$= x^2 + x + y^2 + 6y - 1 = 0$$
$$= x^2 + x + \left(\frac{1}{2}\right)^2 - \left(\frac{1}{2}\right)^2 + y^2 + 6y + 3^2 - 3^2 - 1 = 0$$
$$= x^2 + x + \left(\frac{1}{2}\right)^2 + y^2 + 6y + 3^2 = \left(\frac{1}{2}\right)^2 + 3^2 + 1$$
$$= \left(x + \frac{1}{2}\right)^2 + (y + 3)^2 = \frac{41}{4}$$
$$= \left(x + \frac{1}{2}\right)^2 + (y + 3)^2 = \left(\frac{\sqrt{41}}{2}\right)^2$$

Therefore, the radius is $\frac{\sqrt{41}}{2} \approx 3.202$ units and the center is $(-0.5, -3)$

Example 3

Find the radius and the center of the following circle given by
$$f(x) = 3x^2 + 3y^2 - 12y + 24x + 6 = 0$$

<u>Solution</u>

We collect the terms having similar variable together then complete the square in each case

$$f(x) = 3x^2 + 3y^2 - 12y + 24x + 6 = 3x^2 + 24x + 3y^2 - 12y + 6 = 0$$
$$= 3(x^2 + 8x) + 3(y^2 - 4y) + 6 = 0$$
$$= 3(x^2 + 8x + 4^2 - 4^2) + 3(y^2 - 4y + 2^2 - 2^2) + 6 = 0$$
$$= 3(x^2 + 8x + 4^2) - 48 + 3(y^2 - 4y + 2^2) - 12 + 6 = 0$$
$$= 3(x^2 + 8x + 4^2) + 3(y^2 - 4y + 2^2) = 48 + 6$$
$$= 3(x + 4)^2 + 3(y - 2)^2 = 54 = \left(\sqrt{54}\right)^2 = \left(3\sqrt{6}\right)^2$$

Therefore, the radius is $3\sqrt{6}$ units while the center is $(-4,2)$

Example 4

The find the equation of the circle whose diameter has the end points at $(12, 2)$ and $(-6, 2)$.

<u>Solution</u>

The mid-point of the points $(12, 2)$ and $(-6, 2)$ is the center of the diameter as well as the circle

The center is $\left(\left(\frac{12 + (-6)}{2} \right), \left(\frac{2 + 2}{2} \right) \right) = (3, 2)$

The radius is the distance from the center to the end point of the diameter.

$r = \sqrt{(12 - 3)^2 + (2 - 2)^2} = 9 \; units$

Section 104 Conclusion

In this lesson, we have discussed the concept of limits of functions and saw how limits are used to define continuity of functions at a point as well as on an interval. We have moved further to look at some of the conics, that is, the hyperbola, the parabola and the circle.

CALCULUS: APPLICATION OF CONICS AND INTRODUCTION TO DIFFERENTIATION

Content Description

In this session, we will discuss the ellipse as one of the conics then look at the application of all conics. We will also introduce the concept of differentiation and differentiability.

MATH TOPICS

- Calculus 104.11 Ellipse (Reference 1.11).
- Calculus 104.12 Application of conic sections (Reference 1.12).
- Calculus 104.13 Differentiation, Definition (Reference 1.13).
- Calculus 104.14 Differentiability (Reference 1.14).
- Calculus 104.15 Higher derivatives (Reference 1.15).

INTRODUCTION

All along, the slope is known to be rise divided by run. This is applicable if the graph is a straight line. In situations where the graph is not a straight line for instance a curve or any other graph, we consider other concepts that lead to an exhaustive definition of the term slope. This concept is limits. Due to different types of functions and their nature, a number of methods for determining the slope is provided. We will also look at the last conic as well as their application id day-to-day life.

Section 104.11: Ellipse

An ellipse is a collection of points where the sum of the distances from each point to two fixed points called the focus is the same for all points. For instance, for any point P on the curve, the sum $PF_1 + PF_2$ is equal for all points.

The equation of the ellipse is given by

$$\frac{(x-h)^2}{a^2} + \frac{(y-k)^2}{b^2} = 1$$

Where (h, k) is the center of the ellipse, a and b are the distances from the center to the vertex and co-vertex respectively.

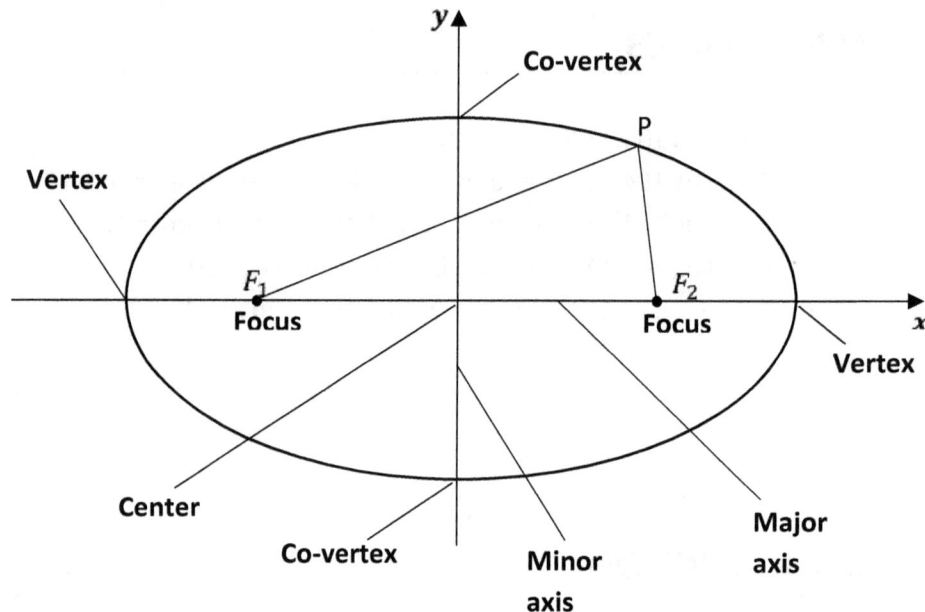

If the major axis is on the $y-$ axis, then the equation becomes

$$\frac{(y-k)^2}{a^2} + \frac{(x-h)^2}{b^2} = 1$$

Where (h, k) is the center of the ellipse, a and b are the distances from the center to the vertex and co-vertex respectively.

The distance from the center to the foci is c, where $c^2 = a^2 - b^2$

Example 1

Find the equation of an ellipse whose foci are $(1,1)$ and $(1,9)$ and the distance from the center to the co-vertex is 2 units.

Solution

The distance from the center to the co-vertex is $b = 2$

The midpoint of the foci is the center and it is given by $\left(\dfrac{1+1}{2}, \dfrac{9+1}{2}\right) = (1,5)$

Since the foci are on the same vertical line, $x = 1$, the ellipse is vertical.

Thus, the distance from the center to the foci, $c = \sqrt{(1-1)+(9-5)^2} = 4$

Thus, we determine a.
$$c^2 = a^2 - b^2 ; a^2 = b^2 + c^2 = 2^2 + 4^2 = 20 ; b^2 = 4$$

Substituting into the equation,
$$\frac{(y-k)^2}{a^2} + \frac{(x-h)^2}{b^2} = 1$$
we get
$$\frac{(y-5)^2}{20} + \frac{(x-1)^2}{4} = 1$$

Example 2

Find the coordinates of the vertices and the foci of the graph $\dfrac{(y+1)^2}{81} + \dfrac{(x-4)^2}{25} = 1$

Solution

The standard equation of the graph would be $\dfrac{(y+1)^2}{9^2} + \dfrac{(x-4)^2}{5^2} = 1$. Comparing with $\dfrac{(y-k)^2}{a^2} + \dfrac{(x-h)^2}{b^2} = 1$, we have;

The center is $(-1,4), a = 9, b = 5$ and the graph is vertical since the intercept for y is more than that of x .

The vertices are 9 units above and below the center. Their coordinates are $(-1, 4+9) = (-1,13)$ and $(-1, 4-9) = (-1,-5)$.

$$c^2 = a^2 - b^2 = 9^2 - 5^2 = 56; \quad c = 7.483 \text{ units}$$

The foci are 7.483 units above and below the center. These are $(-1, 4 + 7.483) = (-1, 11.48)$ and $(-1, 4 - 7.483) = (-1, -3.483)$.

The vertices are $(-1, 13)$ and $(-1, -5)$ while the foci are $(-1, 11.48)$ and $(-1, -3.483)$

Example 3

Find the coordinates of the center and vertices of the graph $\dfrac{(x+2)^2}{64} + \dfrac{(y+1)^2}{16} = 1$

Solution

The standard equation is $\dfrac{(x+2)^2}{8^2} + \dfrac{(x+1)^2}{4^2} = 1$. Comparing with $\dfrac{(x-h)^2}{a^2} + \dfrac{(y-k)^2}{b^2} = 1$, we get;

The center $(-2, -1)$, $a = 8$ and $b = 4$.

The graph is horizontal since the value of the constant below $(x-h)^2$ is more than that below $(y-k)$.

The vertices are 8 units on the left and right of the center. Their coordinates are $(-2, -1 - 8) = (-2, -9)$ and $(-2, -1 + 8) = (-2, 7)$.

Thus, the center is $(-2, -1)$ and the vertices are $(-2, -9)$ and $(-2, 7)$.

Example 4

Find the foci and center of a graph whose vertices are $(-7, -2)$ and $(5, -2)$ and the distance between the co-vertices is 6.

Solution

The center of the vertices is at the intersection of the vertices. Using the midpoint formula, the center is

$$\left(\left(\frac{-7 + 5}{2} \right), \left(\frac{-2 - 2}{2} \right) \right) = (-1, -2)$$

The vertices have the same $y-$ coordinate implying that the ellipse is horizontal

The distance between the vertices is $2a = |5 - -7| = 12; a = 6$

The distance between the co-vertices is $2b = 6; \quad b = 3$.

$$c^2 = a^2 - b^2 = 6^2 - 3^2 = 36 - 9 = 27; \quad c = \sqrt{27} = 3\sqrt{3} \approx 5.196 \text{ units}$$

The foci are 5.196 units to the left and right of the center. Their coordinates are

$(-1-5.196,-2)=(-6.196,-2),(-1+5.196,-2)=(4.196,-2)$

Thus, the foci are $(-6.196,-2)$ and $(4.196,-2)$

Section 104.12: Application of conic sections

Conic sections are widely used in real world. In this lesson, we will highlight a few.

Arches
Most artists use semicircular arches to design the doors and the window openings.

Bridges
A number of bridges are built using parabolic as well as circular arches.

Parabolic mirrors
These mirrors are used in reflectors because the concentrate all light parallel to their line of symmetry at the focus. This makes it too possible to be used in solar heaters

Hyperbolic mirrors
These are the mirrors used as side mirrors for vehicles. This is because they have a wide field of view and their contraction nature of waves. They receive waves from a wide view of directions then concentrate the virtual images at their focus. This makes it possible for people to view objects on a wide view.

Obits
The obits of the planets are elliptic making it easy for scientists to study them in relation to other heavenly bodies.

Example 1
A satellite is launched around the earth's atmosphere to make elliptic revolutions. The distance between the co-vertices of is 8100 miles. The shortest distance between the earth surface and the minor axis is 50 miles while the focus of the vertex is on the surface of the earth a point farthest from the center of the satellite's orbit. Write the equation of the orbit governing the orbit of the satellite if the radius of the earth is approximately 3968 miles. Take the center of the earth as the origin.

Solution

We make a diagram

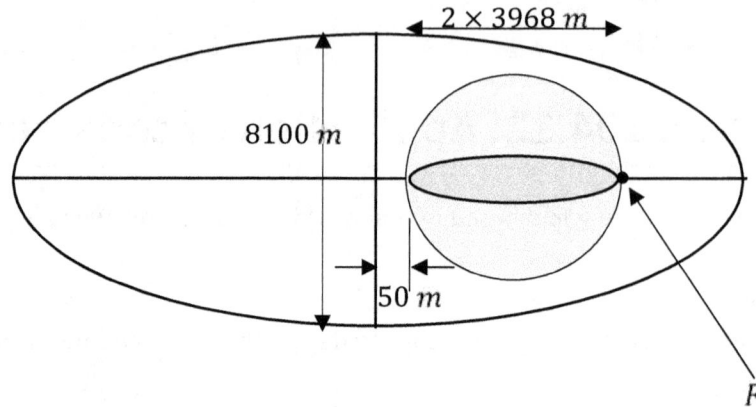

The distance between the co-vertex is $2b = 8100; b = 4050\ mi, b^2 = 16402500$

The focus on the right of the center is at $c = 50\ mi + (2 \times 3968\ mi) = 7986\ mi$

The distance from the center to the vertex is

$a = \sqrt{b^2 + c^2} = \sqrt{4050^2 + 7986^2} = \sqrt{80178696} = 8954$

The center is $50 + 3968 = 4018\ mi$ to the left of the center, though at the same horizontal level.

Thus, the center is $(-8954, 0)$

The equation is $\dfrac{(x + 8954)^2}{80178696} + \dfrac{y^2}{16402500} = 1$

Example 2

A fountain is to be erected some distance away from the building as shown below. It is to be surrounded by a 7 ft thick path. If the lower right-hand side corner of the building is to be taken as the reference point, find the equation describing the boundary of the fountain and that of the path. Form the inequality representing the path.

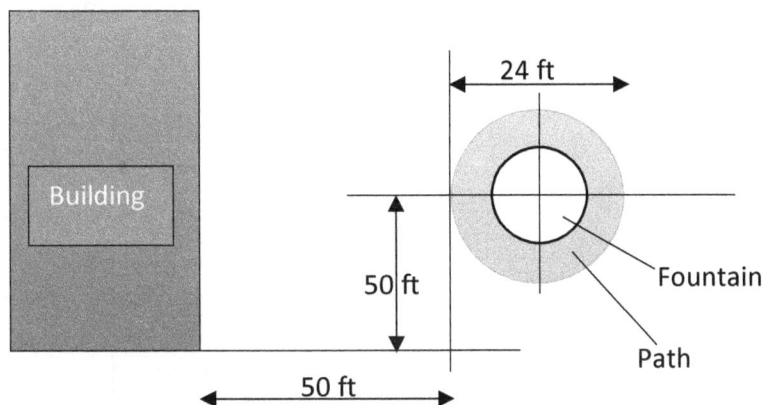

Solution

From reference point, the center is 50 ft upwards and $50 + 12 = 62\ ft$ to the left.

Thus, the center of the boundaries is $(62, 50)$

The radius of the outer boundary, that of the path, is 12 ft while that of the fountain is $12 - 7 = 5\ ft.$

Using the form, $(x - h)^2 + (y - k)^2 = r^2$, we have the equations

That equation of the boundary of the path is $(x - 62)^2 + (y - 50)^2 = 144$ while that of the fountain is $(x - 62)^2 + (y - 50)^2 = 25$.

To describe the earth, we need only to show that r is a variable, that it is varying from 5 to 12.

Thus, the inequality is
$(x - 62)^2 + (y - 50)^2 = r, 5 \leq r \leq 12.$

Example 3

An architect comes up with a drawing having the extract below. If O is taken as the origin of the view of that arch, find the equation of the semicircular arch below.

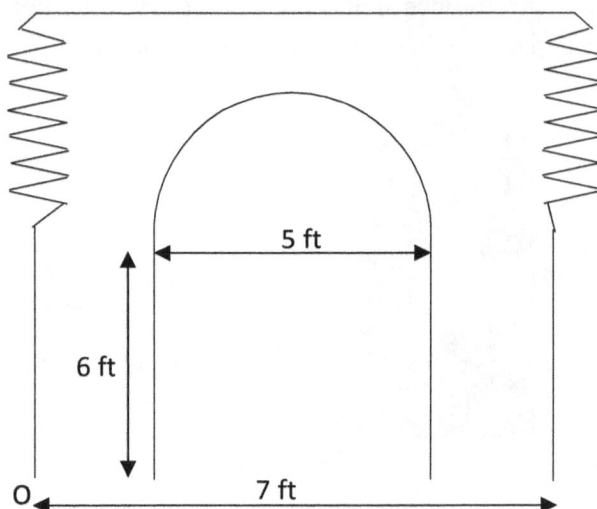

Solution

From the origin, O, the center of the arc is 6 ft vertically and 3.5 ft horizontally. Thus, the center is $(3.5, 6)$.

The radius of the circle is 2.5 ft

The equation of the circle is $(x - 3.5)^2 + (y - 6)^2 = 2.5^2$

Since we are only considering a semi circle, we have to limit what y is so that we can describe the semicircle.

At the lowest point of the circle, the value of y is 6 ft, while at the highest point, we add the radius to get $6 + \dfrac{5}{2} = 8.5$ ft

Thus, the real equation is $(x - 3.5)^2 + (y - 6)^2 = 2.5^2, \qquad 6 \le y \le 8.5$.

Section 104.13 Differentiation, Definition

Consider the slope of the line AB in the following diagram

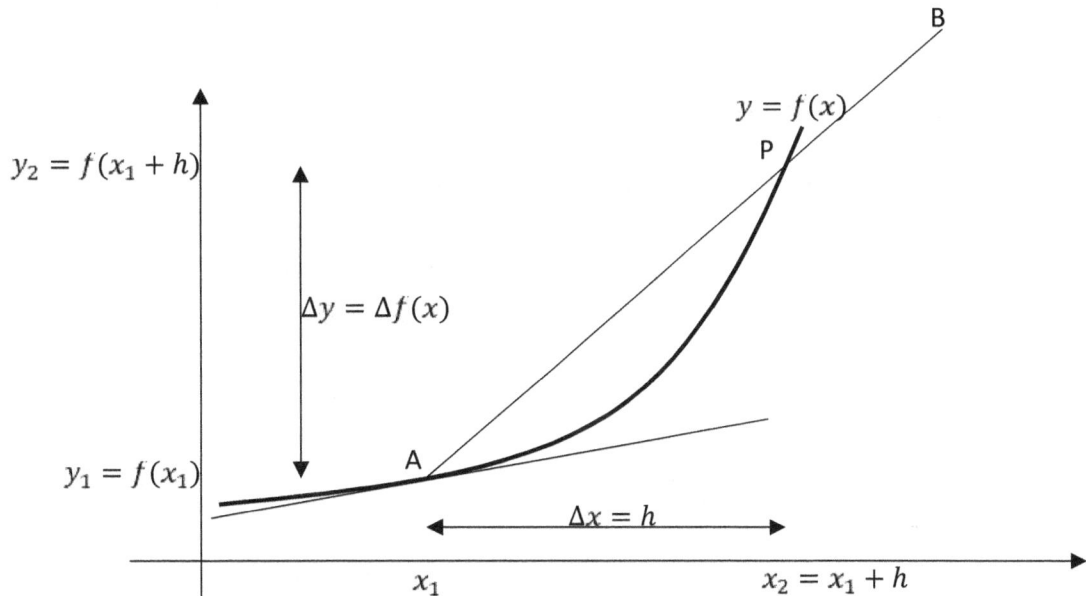

We consider the slope of the point AB, however, our interest is the slope of the tangent at A

The coordinates of A is (x_1, y_1) and that of B is $(x_2, y_2) = (x_1 + h, f(x_1 + h))$

Assume that A is fixed and P is a point on the line AB where the line intersects the curve $y = f(x)$. Thus, P is moving along the come to point A. As it moves, $x_1 + h$ comes closer to x_1 as h decreases in size.

The slope of AB $\quad slope = \dfrac{rise}{run} = \dfrac{change\ in\ y\ or\ f(x)}{Change\ in\ x} = \dfrac{\Delta f}{\Delta x}$ is

Thus, $\quad slope = \dfrac{\Delta f}{\Delta x}$

But $\Delta x = h$ and change in y is the difference in $y-$ values at B and A. That is $f(x_1 + h) - f(x_1)$

Thus the slope is $\quad \dfrac{\Delta f}{\Delta x} = \dfrac{f(x_1 + h) - f(x_1)}{h}$

As B moves down downwards, h reduces. As it approaches to be the tangent at A, h approaches to 0 hence, the slope will be at that instant, instantaneous slope of the curve denoted $\dfrac{df}{dx}$ or $f'(x)$. This is the limit of the slope of AB as h approaches zero.

Thus $$Instantaneous\ slope = \frac{df}{dx} = \lim_{h \to 0} \frac{\Delta f}{\Delta x} = \frac{\lim_{h \to 0} \left(f(x_1 + h) - f(x_1)\right)}{h}$$

This instantaneous slope is called the derivative of the curve. Since we are doing it with respect to the variable x , we say the derivative of $f(x)$ with respect to x .

Thus, derivative of a function $f(x)$ is the limit $\dfrac{\lim_{h \to 0} \left(f(x + h) - f(x)\right)}{h}$ if it exists.

Therefore, differentiation is the process of determining the derivative of the curve with respect to a given variable.

Example 1

What is the derivative of $f(x) = 5x + 1$

Solution

From the definition, the derivative of $f(x)$ is the limit $\dfrac{\lim_{h \to 0} \left(f(x + h) - f(x)\right)}{h}$

$f(x) = 5x + 1$; $f(x + h) = 5(x + h) + 1 = 5x + 5h + 1$

$$\frac{df}{dx} = \frac{\lim_{h \to 0} \left(f(x + h) - f(x)\right)}{h} = \frac{\lim_{h \to 0} \left(5x + 5h + 1 - (5x + 1)\right)}{h} = \frac{\lim_{h \to 0} (5x + 5h + 1 - 5x - 1)}{h}$$

$$= \frac{\lim_{h \to 0} 5h}{h} = \lim_{h \to 0} 5 = 5$$

It can also be confirmed that the slope of the above linearfunction $y = mx + c$ is $m = 5$

Example   **2**

What is the derivative of $f(x) = x^2 + 3x + 1$

Solution

From the definition, the derivative of $f(x)$ is the limit $\dfrac{\lim_{h \to 0} \left(f(x + h) - f(x)\right)}{h}$

$f(x) = x^2 + 3x + 1$;

$f(x + h) = (x + h)^2 + 3(x + h) + 1 = x^2 + 2xh + h^2 + 3x + 3h + 1$

$$\frac{df}{dx} = \frac{\lim_{h \to 0} \left(f(x + h) - f(x)\right)}{h} = \frac{\lim_{h \to 0} \left(x^2 + 2xh + h^2 + 3x + 3h + 1 - x^2 - 3x - 1\right)}{h}$$

$$= \frac{\lim\limits_{h \to 0} (2xh + h^2 + 3h)}{h} = \frac{\lim\limits_{h \to 0} (2x + h + 3)h}{h} = \lim\limits_{h \to 0} 2x + h + 3$$

$$= 2x + 0 + 3 = 2x + 3$$

$$Thus, \frac{df}{dx} = 2x + 3$$

Example 3

What is the derivative of $f(x) = \dfrac{1}{4x}$

Solution

From the definition, the derivative of $f(x)$ is the limit $\dfrac{\lim\limits_{h \to 0} \left(f(x+h) - f(x)\right)}{h}$

$$f(x) = \frac{1}{4x}; \quad f(x+h) = \frac{1}{4(x+h)} = \frac{1}{4x + 4h}$$

$$\frac{\lim\limits_{h \to 0} \left(f(x+h) - f(x)\right)}{h} = \lim\limits_{h \to 0} \left(\frac{1}{4x + 4h} - \frac{1}{4x}\right)\frac{1}{h} = \lim\limits_{h \to 0} \left(\frac{4x - 4x - 4h}{4x(4x + 4h)}\right)\frac{1}{h}$$

$$= \lim\limits_{h \to 0} \left(\frac{-4h}{4x(4x + 4h)}\right)\frac{1}{h} = \lim\limits_{h \to 0} \left(\frac{-4}{4x(4x + 4h)}\right)$$

$$= \lim\limits_{h \to 0} \left(\frac{-4}{4x(4x + 4h)}\right) = \frac{-4}{4x(4x + 4(0))} = -\frac{4}{16x^2} = -\frac{1}{4x^2}$$

Example 3

What is the derivative of $f(x) = \dfrac{x - 1}{2x + 3}$

Solution

From the definition, the derivative of $f(x)$ is the limit $\dfrac{\lim\limits_{h \to 0} \left(f(x+h) - f(x)\right)}{h}$

$$f(x) = \frac{x - 1}{2x + 3}; \quad f(x+h) = \frac{x + h - 1}{2(x + h) + 3} = \frac{x + h - 1}{2x + 2h + 3}$$

Therefore,

$$\lim_{h \to 0} \frac{\square \left(f(x+h) - f(x) \right)}{h} = \lim_{h \to 0} \left(\frac{x+h-1}{2x+2h+3} - \frac{x-1}{2x+3} \right) \frac{1}{h}$$

$$= \lim_{h \to 0} \left(\frac{x+h-1}{2x+2h+3} - \frac{x-1}{2x+3} \right) \frac{1}{h} = \lim_{h \to 0} \left(\frac{(2x+3)(x+h-1) - (x-1)(2x+2h+3)}{(2x+2h+3)(2x+3)} \right) \frac{1}{h}$$

$$= \lim{}_\top (h \to 0) \; ((2x^\top 2 + 2xh - 2x + 3x + 3h - 3 - 2x^\top 2 - 2xh - 3x + 2x + 2h + 3))/((2x+2h+3)(2x+3))) \; 1$$

$$= \lim_{h \to 0} \left(\frac{2xh + 5h - 2xh}{(2x+2h+3)(2x+3)} \right) \frac{1}{h} = \lim_{h \to 0} \left(\frac{h(2x+5-2x)}{(2x+2h+3)(2x+3)} \right) \frac{1}{h}$$

$$= \lim_{h \to 0} \left(\frac{5}{(2x+2h+3)(2x+3)} \right)$$

To evaluate the limits, we substitute h with zero

$$= \left(\frac{5}{(2x+2(0)+3)(2x+3)} \right) = \frac{5}{(2x+3)^2}$$

Thus,

$$\frac{df}{dx} = \frac{5}{(2x+3)^2}$$

Section 104.14: Differentiability

The concept of differentiability is derived from the derivative of a function. Differentiability 'asks a question' is a function differentiable (at a point) or not? A function is differentiable at a point if the derivative exists at that point. If the derivative does not exist, then we have a situation $\frac{df}{dx} = \infty$, which implies that function has a vertical tangent, a cusp, a break, a jump that point.

Therefore, for a function to be differentiable, it has to be relatively smooth at a point. This

The figure below shows images of cusps, breaks, sharp turn, and jumps that makes a function not differentiable at a point.

Cusps

sharp turn

Jump

break

Example 1

Discuss the continuity and differentiability of $f(x) = 2|x + 1| - 2$ at $x = -1$.

<u>Solution</u>

The function has a vertex at (h, k) given the general equation $f(x) = a|x - h| + k$. Thus, $(h, k) = (-1, -2)$.

By definition of absolute value functions
$$f(x) = \begin{cases} -2|x + 1| - 2 \; ; \; x < -1 \\ 2|x + 1| - 2; \quad x \geq -1 \end{cases}$$

For $f(-1) = 2|-1 + 1| - 2 = -2$, the function is defined

Left hand side limit

$$\lim_{x \to -1^-} -2|x+1| - 2 = -2(0) - 2 = -2$$

Right hand side limit

$$\lim_{x \to -1^+} 2|x+1| - 2 = 2(0) - 2 = -2$$

The limits are equal, hence

$$\lim_{x \to -1} -2|x+1| - 2 = -2(0) - 2 = -2 = f(-2)$$

This implies that the function is continuous.

Differentiability

For $x \geq -1$, $f(x)$ has a positive slope while at $x < -1$, the function has a negative slope. Between and after $x = -1$, the function turns sharply such that it has a vertical tangent at $x = -1$. Thus, the function is nor differentiable at $x = -1$.

Example 2

Is the function $x^7 - y^2 = 0$ differentiable at $x = 0$?

Solution

The function has a cusp at $x = 0$. Thus, at that point, it has a vertical tangent whose slope is not defined. Thus, the function is not differentiable at $x = 0$.

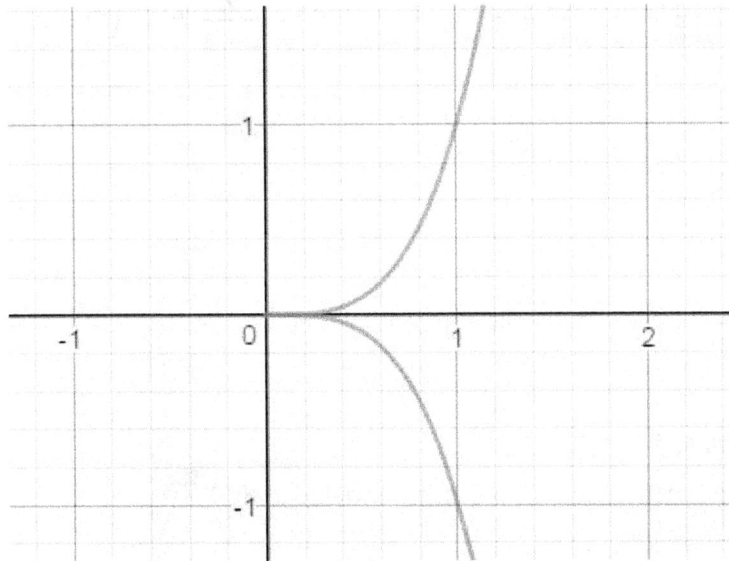

Example 3

Is the function $f(x) = \begin{cases} 2x, & x > 1 \\ 3, & x \leq 1 \end{cases}$ differentiable at $x = 1$?

Solution

$x = 1$ is in the interval $x \leq 1$, hence $f(1) = 3$

$$\frac{df}{dx} = \lim_{h \to 0} \frac{f(x+h) - f(x)}{h}$$

Let x be a general point approaching 1, hence $h = x - a$

Thus, we have $h \to 0$ implying $x - a \to 0$, or $x \to a$

Thus, the function changes to

$$\frac{df}{dx} = \lim_{x \to a} \frac{f(x) - f(a)}{x - a}$$

We consider sided limits

Left hand side limit

$$L = \lim_{x \to 1^-} \frac{f(x) - f(1)}{x - 1} = \lim_{x \to 1^-} \frac{2x - 3}{x - 1} = \frac{2(1) - 3}{1 - 1} = -\frac{1}{0} \; undefined$$

Right hand side limit

$$R = \lim_{x \to 1^+} \frac{f(x) - f(1)}{x - 1} = \lim_{x \to 1^-} \frac{3 - 3}{x - 1} = \lim_{x \to 1^-} \frac{0}{x - 1} = \lim_{x \to 1^-} 0 = 0$$

Since $L \neq R$, the limit $\lim_{x \to 1} \frac{f(x) - f(1)}{x - 1}$

Does not exists.

Thus, the function is not differentiable at $x = 1.$

Example 4

Is the function $g(x) = \sqrt{x}$ differentiable at $x = 0$?

<u>Solution</u>

From the definition, the function is differentiable if the limit

$$\frac{df}{dx} \; (at \; x = a) = \lim_{x \to a} \frac{f(x) - f(a)}{x - a} \; exists.$$

Since $x = 0$ is in the domain of the function, it is continuous at that function, hence, we proceed with determining if the derivative exists at that point.

$$\frac{df}{dx} \; (at \; x = a) = \lim_{x \to 0} \frac{\sqrt{x} - \sqrt{0}}{x - 0} = \lim_{x \to 0} \frac{\sqrt{x}}{x} = \lim_{x \to 0} \frac{1}{\sqrt{x}} = \infty$$

Hence, the limit does not exists.

Section 104.15: Higher derivatives

Derivatives of a function can be determined again and gain so long as the answer got after each step is differentiable. When each of the derivative of a function is continuous, and the derivative can be done infinite number of times, we say the function is continuously differentiable. If we carry out a derivative of a function once, we call it the first derivative.

If we differentiate the first derivative, that is differentiating the first function twice, we call it a second derivative. Under the same scenario, we can come up with third derivative, fourth derivatives among others. Given the function $f(x) = y$,

the first derivative is $y' = \dfrac{df}{dx}$

The second derivative is $y'' = \dfrac{d}{dx}\left(\dfrac{df}{dx}\right) = \dfrac{d^2 f}{dx^2}$

The third derivative is $y''' = \dfrac{d}{dx}\left(\dfrac{d^2 f}{dx^2}\right) = \dfrac{d^3 f}{dx^3}$

The Fourth derivative is $y^{IV} = \dfrac{d}{dx}\left(\dfrac{d^3 f}{dx^3}\right) = \dfrac{d^{IV} f}{dx^{IV}}$

And so on

We also not that, using differentiation from the first principle, it can be proved that the derivative of $f(x) = ax^n$ where a and n are constants with $a \neq 0$, can be given as $\dfrac{df}{dx} = nax^{n-1}$.

If $f(x) = c$ where c is a constants, then $\dfrac{df}{dx} = 0$.

Example 1

Find the first derivative of $f(x) = 3x^3 + \dfrac{1}{3}x^2 + 6x + 1$

<u>Solution</u>

$f(x) = 3x^3 + \dfrac{1}{3}x^2 + 6x + 1$

$\dfrac{df}{dx} = \dfrac{d}{dx}3x^3 + \dfrac{d}{dx}\dfrac{1}{3}x^2 + \dfrac{d}{dx}6x + \dfrac{d}{dx}1 = 9x^{3-1} + \dfrac{2}{3}x^{2-1} + 6x^{1-1} + 0 = 9x^2 + \dfrac{2}{3}x + 6$

Hence $\dfrac{df}{dx} = 9x^2 + \dfrac{2}{3}x + 6$

Example 2

Find the first and second derivative of $f(x) = 3x^7 + \dfrac{1}{2}x^5 + 6x^2 + 1$ at $x = 1$.

Solution

$$f(x) = 3x^7 + \frac{1}{2}x^5 + 6x^2 + 1$$

We apply the method $\dfrac{df}{dx} = nax^{n-1}$ on each term to get

$$\frac{df}{dx} = 21x^{7-1} + \frac{5}{2}x^{5-1} + 12x^{2-1} + 0 = 21x^6 + \frac{5}{2}x^4 + 12x$$

$$\frac{df}{dx} = 21x^6 + \frac{5}{2}x^4 + 12x$$

At $x = 1$, the derivative is

$$\frac{df}{dx}\bigg|_{x=1} = 21(1)^6 + \frac{5}{2}(1)^4 + 12(1) = 21 + \frac{5}{2} + 12 = 35.5$$

To get the second derivative, we differentiate the first derivative

$$y'' = \frac{d^2f}{dx^2} = \frac{d}{dx}\left(\frac{df}{dx}\right) = \frac{d}{dx}\left(21x^6 + \frac{5}{2}x^4 + 12x\right) = 126x^{6-1} + 10x^{4-1} + 12x^{1-1}$$

Thus, the second derivative is $\dfrac{d^2f}{dx^2} = 126x^5 + 10x^3 + 12$

At $x = 1$, the second derivative is

$$\frac{d^2f}{dx^2}\bigg|_{x=1} = 126(1)^5 + 10(1)^3 + 12 = 126 + 10 + 12 = 148$$

Example 3

Find the first, second and third derivative of $f(x) = 6x^2 + 5x + 1$.

Solution

We apply the method $\dfrac{df}{dx} = nax^{n-1}$ on each term to get

$$\frac{df}{dx} = \frac{d}{dx}(6x^2 + 5x + 1) = 12x^{2-1} + 5x^{1-1} + 0 = 12x + 5$$

To get the second derivative, we differentiate the first derivative once

$$y'' = \frac{d^2f}{dx^2} = \frac{d}{dx}\left(\frac{df}{dx}\right) = \frac{d}{dx}(12x + 5) = 12x^{1-1} + 0 = 12$$

$$\frac{d^2f}{dx^2} = 12$$

To get the third derivative, we differentiate the second derivative once

$$y''' = \frac{d^3f}{dx^3} = \frac{d}{dx}\left(\frac{d^2f}{dx^2}\right) = \frac{d}{dx}12 = 0$$

Thus,

$$\frac{d^3f}{dx^3} = 0$$

Section 104 Conclusion

In this lesson, we have discussed the equation of the ellipse and seen how conic sections can be applied in real life situations. We have also introduced the concept of differentiation by defining it and determining the conditions under which a function can be differentiable. We have complete the lesson by looking at how of determine higher derivative of functions.

CALCULUS: APPLICATION OF CONICS AND INTRODUCTION TO DIFFERENTIATION

Content Description

In this session, we will discuss different rules used in differentiation as well the application of differentiation in curve sketching and optimization. We also look at the main theorems in differentiation.

MATH TOPICS

- Calculus 104.16 Rules of differentiation (Reference 1.11).
- Calculus 104.17 Curve Sketching (Reference 1.12).
- Calculus 104.18 Optimization (Reference 1.13).
- Calculus 104.19 The Rolles Theorem (Reference 1.14).
- Calculus 104.20 Mean Value Theorem (Reference 1.15).

INTRODUCTION

There are different types of functions that we meet in both real and academic life; the rate of these functions is referred to as the derivative. Due to different types of function, the basic method of differentiation learnt is not enough to different them in good time, we need some specific viable methods. That is why we need to look at rules of differentiation. As an extension of the concept, we will also look at the application of differentiation as well as some results on differentiation that describes some functions.

Section 104.16: Rules of differentiation

These are rules that enable us carry out differentiation of different types of functions.

Derivative of a constant

The derivative of a $y = f(x) = c$ where c is a constant is $\dfrac{dy}{dx} = 0$

Addition and subtraction

If $y = f(x) \pm h(x)$, then $\dfrac{dy}{dx} = \dfrac{df}{dx} \pm \dfrac{dh}{dx}$

Multiplication by a constant

If $y = kf(x)$ then $\dfrac{dy}{dx} = k\dfrac{df}{dx}$

Product rule

If $y(x) = f(x)g(x)$ then $\dfrac{dy}{dx} = \dfrac{df}{dx}g(x) + \dfrac{dg}{dx}f(x)$

Quotient rule

If $y(x) = \dfrac{f(x)}{g(x)}$ then

$$\frac{dy}{dx} = \frac{\dfrac{df}{dx}g(x) - \dfrac{dg}{dx}f(x)}{[g(x)]^2}$$

Power rule

If $y(x) = (f(x))^n$ where n is constant, then

$$\frac{dy}{dx} = n(f(x))^{n-1}\frac{df}{dx}$$

Chain rule

If $y(x) = f(g(x))$ then

$$\frac{dy}{dx} = f'(g(x))g'(x) = \frac{df}{dg}\frac{dg}{dx}$$

Exponential differentiation

If $y(x) = e^{u(x)}$, then

$$\frac{dy}{dx} = u'(x)e^{u(x)}$$

If $y(x) = a^{u(x)}$, where a is any other constant other than e then

$$\frac{dy}{dx} = u'(x)e^{u(x)}\log a$$

Differentiation of logarithms

If $y = \ln(u(x))$ then

$$\frac{dy}{dx} = \frac{u'(x)}{u(x)}$$

Trigonometric differentiation

$$y = \sin x, y' = \cos x$$
$$y = \cos x, \qquad y' = -\sin x$$
$$y = \tan x, y' = \sec^2 x$$
$$y = \cot x; \quad y' = -\csc x$$
$$y = \sec x, y' = \sec x \tan x$$
$$y = \csc x, y' = -\csc x \cot x$$

Example 1

Find the derivative of the following functions

(i) $\quad y = \dfrac{2x^2}{4x - 8}$

(ii) $\quad y = (3x + 4)^3(4x - 9)$

(iii) $\quad y = \sin^2(2x + 3)$

(iv) $\quad y = e^{4x + \tan\left(\frac{x}{2}\right)}$

(v) $\quad y = x\ln(3x + 1)$

(vi) $\quad y = \sec\left(e^{(2x-1)^2}\right)$

Solution

(i). We use the quotient rule

$$y = \frac{2x^2}{4x - 8} = \frac{f(x)}{g(x)}$$

Then

$$\frac{dy}{dx} = \frac{\frac{df}{dx}g(x) - \frac{dg}{dx}f(x)}{[g(x)]^2} = \frac{(4x)(4x-8) - 4(2x^2)}{(4x-8)^2} = \frac{16x^2 - 32x - 8x^2}{(4x-8)^2}$$

$$= \frac{8x^2 - 32x}{(4x-8)^2} = \frac{8x(x-4)}{(4(x-2))^2} = \frac{8x(x-4)}{16(x-2)^2} = \frac{x(x-4)}{2(x-2)^2}$$

Hence $\dfrac{dy}{dx} = \dfrac{x(x-4)}{2(x-2)^2}$

(ii) We first use product rule, if $y(x) = f(x)g(x)$ then $\dfrac{dy}{dx} = \dfrac{df}{dx}g(x) + \dfrac{dg}{dx}f(x)$

$$y = (3x+4)^3(4x-9)$$

Let $f(x) = (3x+4)^3$ and $g(x) = 4x - 9$

Using power rule $on\ (3x+4)^3, \dfrac{df}{dx} = 3(3x+4)^2(3) = 9(3x+4)^2; \dfrac{dg}{dx} = 4$

Thus,

$$\frac{dy}{dx} = \frac{df}{dx}g(x) + \frac{dg}{dx}f(x) = 9(3x+4)^2(4x-9) + 4(3x+4)^3$$

Upon factorization, we get
$$(3x+4)^2(9(4x-9) + 4(3x+4)) = (3x+4)^2(36x - 81 + 12x + 16)$$
$$= (3x+4)^2(48x - 97)$$

Therefore, $\dfrac{dy}{dx} = (3x+4)^2(48x - 97)$

(iii). $y = \sin^2(2x + 3)$

We use chain rule. Let $h = h(x) = 2x + 3$, then $y = \sin^2 h = (\sin h)^2$

Thus, $\dfrac{dh}{dx} = 2$ and by power rule $\dfrac{dy}{dh} = 2\sin h \cos h$

Thus, $\dfrac{dy}{dx} = \dfrac{dy}{dh} \times \dfrac{dh}{dx} = 2(2\sin h \cos h)$

Substituting for h , we get

$$\dfrac{dy}{dx} = 4\sin(2x + 3)\cos(2x + 3)$$

(iv). $y = e^{4x + \tan\left(\frac{x}{2}\right)}$

We use chain rule. Let $h(x) = 4x + \tan\left(\dfrac{x}{2}\right)$, then $y = e^h$

$$\dfrac{dh}{dx} = 4 + \dfrac{d}{dx}\left(\tan\left(\dfrac{1}{2}x\right)\right) = 4 + \dfrac{1}{2}\tan\left(\dfrac{1}{2}x\right)\sec^2\left(\dfrac{1}{2}x\right)$$

$$\dfrac{dy}{dh} = e^h$$

Thus, by chain rule,

$$\dfrac{dy}{dx} = \dfrac{dy}{dh}\dfrac{dh}{dx} = e^h\left(4 + \dfrac{1}{2}\tan\left(\dfrac{1}{2}x\right)\sec^2\left(\dfrac{1}{2}x\right)\right) = e^{4x + \tan\left(\frac{x}{2}\right)}\left(4 + \dfrac{1}{2}\tan\left(\dfrac{1}{2}x\right)\sec^2\left(\dfrac{1}{2}x\right)\right)$$

Hence,

$$\dfrac{dy}{dx} = e^{4x + \tan\left(\frac{x}{2}\right)}\left(4 + \dfrac{1}{2}\tan\left(\dfrac{1}{2}x\right)\sec^2\left(\dfrac{1}{2}x\right)\right)$$

(v). $y = x\ln(3x + 1)$

We use product rule, let $y = f(x)g(x)$ so that $f(x) = x$ and $= g(x) = \ln(3x + 1)$

We use the rule of differentiating functions with natural logarithms

$f'(x) = 1$ and $g'(x) = \dfrac{3}{3x + 1}$

Thus,

$$\dfrac{dy}{dx} = f'(x)g(x) + f(x)g'(x) = 1(\ln(3x + 1)) + \dfrac{3}{3x + 1}(\ln(3x + 1))$$

$$= \ln(3x + 1)\left(1 + \dfrac{3}{3x + 1}\right) = \ln(3x + 1)\left(\dfrac{3x + 1 + 3}{3x + 1}\right) = \ln(3x + 1)\left(\dfrac{3x + 4}{3x + 1}\right)$$

Hence, $\dfrac{dy}{dx} = \left(\dfrac{3x+4}{3x+1}\right)\ln(3x+1)$

(vi). $y = \sec\left(e^{(2x-1)^2}\right)$

We use chain rule

Let $u = e^{(2x-1)^2}$, $y = \sec u$

$$\frac{du}{dx} = e^{(2x-1)^2}\frac{d}{dx}(2x-1)^2 = e^{(2x-1)^2}2(2x-1)(2) = 4(2x-1)e^{(2x-1)^2}$$

$$\frac{dy}{du} = \sec u \tan u$$

Thus,
$$\frac{dy}{dx} = \frac{dy}{du} \times \frac{du}{dx} = \sec u \tan u\, 4(2x-1)e^{(2x-1)^2}$$

Upon substitution, we get
$$\frac{dy}{dx} = \sec\left(e^{(2x-1)^2}\right)\tan\left(e^{(2x-1)^2}\right)4(2x-1)e^{(2x-1)^2}$$

Section 104.16: Curve Sketching

Curve sketching is done when the key features of a curve are identified. These features are the intercepts and the turning points.

The intercepts

$x-$ intercept; it is a point where the graph intersects the $x-$ axis. At this point, the $y-$ coordinate is zero.

$y-$ intercept; it is a point where the graph intersects the $y-$ axis. At this point, the $x-$ coordinate is zero.

Turning points

These are points at which the slope of the graph changes. At these points, the slope is zero. These points are maxima, minima and point of inflection.

Maxima

A function $y = f(x)$ has a maxima at $x = a$ is $f'(a) = 0$ and $f''(a) < 0$.

That means around $x = a$, the slope to the left is positive and that to the right is negative.

Minima

A function $y = f(x)$ has a minima at $x = a$ is $f'(a) = 0$ and $f''(a) > 0$.

That means around $x = a$, the slope to the left is negative and that to the right is positive.

Point of inflection

This is a point where the curvature of the curve changes. This may be from concave upwards to downwards or vice verse. If the point of inflection is $x = a$, then $f'(a) = f''(a) = 0$.

To do curve sketching, we have to take not of all these points. We will use examples as illustrations.

Example 1

Sketch the graphs of the following function

(i). $y = 4x^2 - 25$

(ii). $y = 2x^3 + 3x^2 - 9x - 10$

(iii). $y = 2x^3 - 3x^2 - 2x + 3$

Solution

(i). $y = 4x^2 - 25$

$x -$ intercept

At this point, $y = 0$, hence $4x^2 - 25 = 0$; $x^2 = \dfrac{25}{4} = 6.25$, $x = \sqrt{6.25} = \pm 2.5\ in$

$y -$ intercept

At this point, $x = 0$, hence $y = -25$

Maxima or minima

The function has a turning point when $y' = 0$, $y'(x) = 8x = 0$, thus $x = 0$

$y''(x) = 8 > 0$, thus, the turning point is a minima

when $x = 0, y = -25$

The minima is at $(0, -25)$. This also the $y -$ intercept of the function.

Thus, the key features are

$x-intercepts$ at $(2.5,0)$ and $(-2.5,0)$, $y-$ intercept and minima both at $(0,-25)$

Skecting the graph based on these, we have

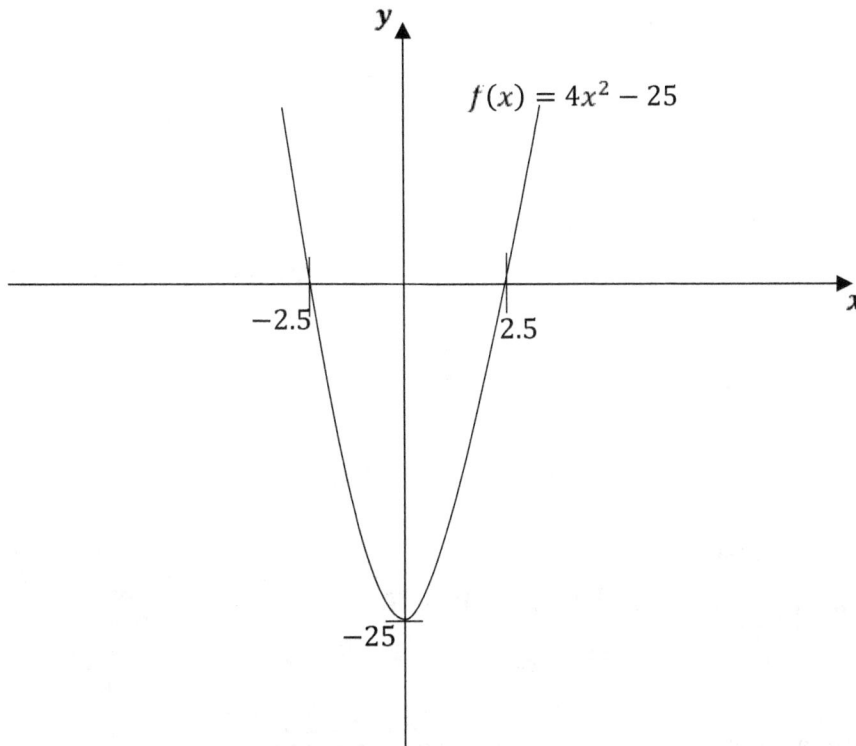

$$f(x) = 4x^2 - 25$$

(ii). $y = 2x^3 + 3x^2 - 9x - 10$

$y-$ intercept

At this point, $x = 0$, thus $y(0) = 2(0) + 3(0) - 9(0) - 10 = -10$

The point IS $(0,-10)$

$x-$ intercept

At this point, $y = 0$, thus, we solve $2x^3 + 3x^2 - 9x - 10 = 0$

Using rational zero theorem, the roots may be $\pm 1, \pm 2, \mp 5, \mp 10, \mp\dfrac{1}{2}, \mp\dfrac{5}{2}.$

By try 2

$y(2) = 2(2^3) + 3(2^2) - 9(2) - 10 = 0$

Thus, $x = 2$ is a solution implying that $x - 2$ is factor of the polynomial

Using synthetic division, we have

$$\begin{array}{r|rrrr} 2 & 2 & 3 & -9 & -10 \\ 0 & 4 & 14 & 10 \\ \hline & 2 & 7 & 5 & 0 \end{array}$$

Thus, when $x - 2$ divides $2x^3 + 3x^2 - 9x - 10$, we get $2x^2 + 7x + 5$

We again factorize the expression
$$2x^2 + 7x + 5 = 2x^2 + 5x + 2x + 5 = x(2x + 5) + 1(2x + 5) = (x + 1)(2x + 5)$$

Thus, we have $x = -1, \qquad x = -\dfrac{5}{2} = -2.5$

The x intercepts are $x = 2, x = -1$ and $x = -2.5$

Turning points
The function is $y = 2x^3 + 3x^2 - 9x - 10$
$$y'(x) = 6x^2 + 6x - 9 = 0 \quad \text{or} \quad 2x^2 + 2x - 3 = 0$$

By quadratic formula $\quad x = \dfrac{-2 \pm \sqrt{28}}{4} = \dfrac{-2 \pm 5.292}{4}$

$x = 0.823, x = -1.823$
For $x = 0.823, y = -14.26$
$\quad x = -1.823, y = 4.26$

Second derivative test.
$$y''(x) = 12x + 6$$

For $x = 0.823, y'' = 15.88 > 0 \quad$ Minima

For $x = -1.823, y'' = -15.88 < 0 \quad$ Maxima

Point of inflection
$$y''(x) = 12x + 6 = 0; x = 0.5$$
At $x = 0.5, \qquad y = -13.5$

The points are

Intercepts, $(0, -10), (2, 0), (-1, 0)$ and $(-2.5, 0)$

Maxima $(-1.823, 4.26)$

Minima $(0.823, -14.26)$

Point of inflection $(0.5, -13.5)$

The sketch is

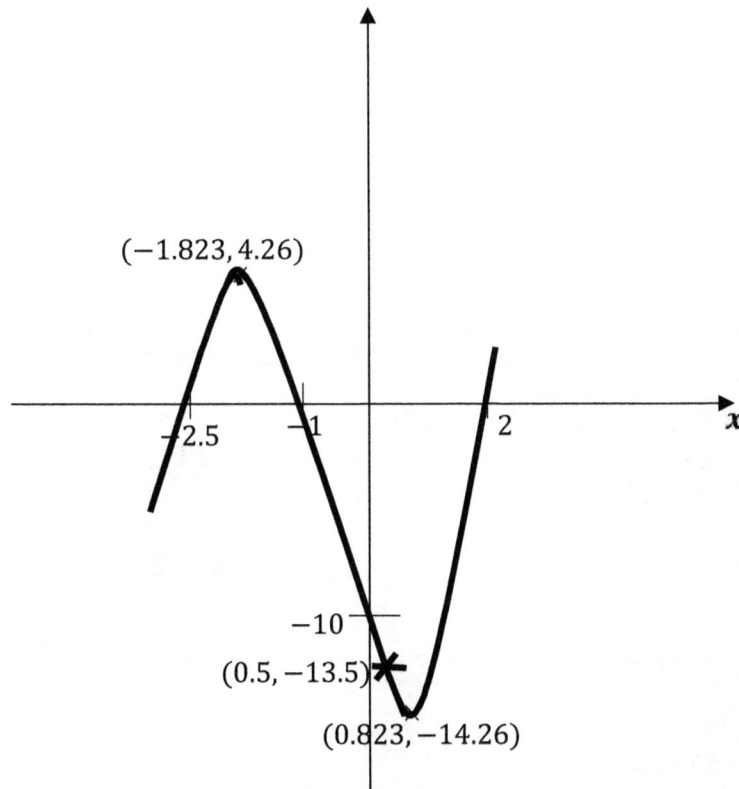

(iii). $y = 2x^3 - 3x^2 - 2x + 3$

Solution

$x -$ intercept

At $x -$ intercept, $y = 0$, hence $2x^3 - 3x^2 - 2x + 3 = 0$. By rational zero theorem, $\pm 1, \pm 3 \; or \; \pm \dfrac{3}{2}$ is a solution of the polynomial.

$y(1) = 2(1) - 3(1) - 2(1) + 3 = 0$, thus, $x = 1$ is solution implying $x - 1$ is a factor of the polynomial.

Using the direct synthesis method, we get

$$
\begin{array}{r|rrrr}
1 & 2 & -3 & -2 & 3 \\
\underline{\quad 2 \quad -1 \quad -3 \quad} & & & & \\
& 2 & -1 & -3 & 0
\end{array}
$$

The quotient is $2x^2 - x - 3$

Factorizing the expression, we get
$$2x^2 - x - 3 = 2x^2 - 3x + 2x - 3 = x(2x - 3) + 1(2x - 3) = (x + 1)(2x - 3)$$

$Thus, x + 1 = 0 \; and \; 2x - 3 = 0$ implies that $x = -1$ and $x = \dfrac{3}{2} = 1.5$

The $x -$ intercepts are $x = 1, x = -1$ and $x = 1.5$

$y -$ intrercept

This occurs when $x = 0$, $y = 2(0) - 3(0) - 2(0) + 3 = 3$

The turning points

At the turning points $y'(x) = 0$.

$y'(x) = 6x^2 - 6x - 2 = 0; \quad 3x^2 - 3x - 1 = 0$

Using quadratic formula,
$$x = \frac{3 \pm \sqrt{21}}{6}; \quad x = -0.2638; \quad x = 1.264$$

Second derivative test

$y''(x) = 12x - 6$

$y''(-0.2638) = 12(-0.2638) - 6 < 0$ maxima

$y''(1.264) = 12(1.264) - 6 > 0$ minima

When $x = -0.2638,$ $y = 3.282$

$x = 1.264,$ $y = -0.2821$

Point of inflection

$y''(x) = 12x - 6 = 0;$ $x = 0.5$

When $x = 0.5, y = 1.5$

The points are

Intercepts, $(0,3), (1,0), (-1,0)$ and $(1.5,0)$

Maxima $(-0.2638, 3.282)$

Minima $(1.264, -0.2821)$

Point of inflection $(0.5, 1.5)$

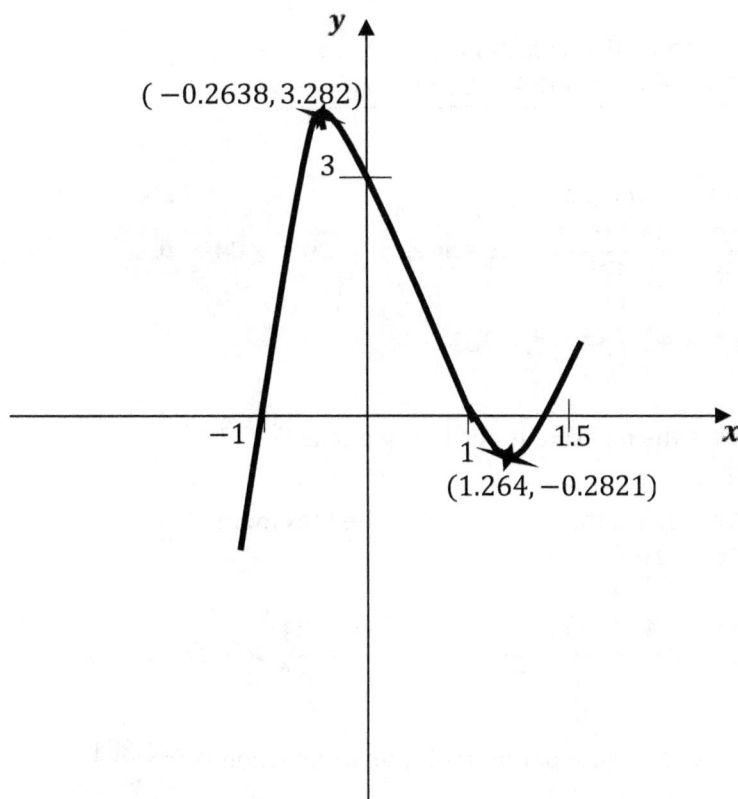

Section 104.18 Optimization

This refers to maximization and minimization of functions. We will use the concepts of maximizing and minimizing functions to optimize variables in the application world.

Some of the most common measures that are optimized include, cost, profit, loss, area, distance or perimeter, volume among others.

We will approach the discussion using examples

Example 1

Minimize the distance around a rectangular piece of land whose area is supposed to be 42 sq. ft

<u>Solution</u>

The distance around the piece of land is perimeter $p = 2(l + w)$

Are is given by $a = lw = 42$; thus $l = \dfrac{42}{w}$

Upon substitution, perimeter would be

$$p(w) = 2\left(\frac{42}{w} + w\right) = \frac{2(42 + w^2)}{w} = \frac{84 + 2w^2}{w}$$

Differentiating with respect to w , we get

$$\frac{dp}{dw} = \frac{4w(w) - (84 + 2w^2)}{w^2} = \frac{2w^2 - 84}{w^2}$$

At the turning point,

$$\frac{dp}{dw} = \frac{2w^2 - 84}{w^2} = 0; \; thus, \qquad 2w^2 - 84 = 0$$

$$w^2 = 42; \quad w = \pm 6.481$$

Since the measure is positive, we take $w = 6.481$

We confirm that this would be the maximum

$$\frac{dp}{dw} = \frac{2w^2 - 84}{w^2}$$

$$\frac{d^2p}{dw^2} = \frac{4w(w^2) - 2w(2w^2 - 84)}{w^4} = \frac{84}{w^4} > 0 \; for \; w > 0$$

Thus, the width perimeter is minimum when $w = 6.381$

The length is $l = \dfrac{42}{w} = \dfrac{42}{6.381} = 6.582\ ft$

Thus, the piece of land measures $6.582\ ft\ by\ 6.481\ ft$

Example 2

A farmer would like to fence her rectangular farm with a 500 feet long wire. Determine the maximum area that she can fence.

Solution

Perimeter is given by $p = 2(l + w)$ where l and w are length and width respectively

Area $a = lw$

We are given that $2(l + w) = 500$, or $l + w = 250$. This implies that $l = 250 - w$

Substituting into the area, we get $a(w) = w(250 - w) = 250w - w^2$

To maximize area, we use the concept of turning points

$\dfrac{da}{dw} = 250 - 2w = 0,$ at the turing point

Thus, $2w = 250;\ \ w = 125ft$

Second derivative test

$\dfrac{d^2a}{dw^2} = -2,\ thus,\ \ \ \ w = 125ft$ is at the maximum

When $w = 125\ ft,\ \ \ \ l = 250 - 125 = 125\ ft$

Thus, the maximum area is at $w = 125ft$. From $a = 250w - w^2$, we have

$a_{max} = 250(125) - 125^2 = 15625\ sq.ft$

Example 3

A water tank of capacity 324 cubic feet is to be constructed using the smallest possible amount of material. What would be the radius of this tank?

Solution

The material represents the surface area. Thus, we would like to minimize the surface area (e).

The surface area is given by $s = 2\pi r^2 + 2\pi rh$

We make s a function of one variable, r, hence, we eliminate h

The volume is $v = \pi r^2 h = 324;\ \ h = \dfrac{162}{\pi r^2}$

Upon substitution, we have

$$s(r) = 2\pi r^2 + 2\pi r \left(\frac{162}{\pi r^2}\right) = 2\pi r^2 + \frac{324}{r} = \frac{2\pi r^3 + 324}{r}$$

Upon differentiating[s] , we get

$$\frac{ds}{dr} = \frac{6\pi r^3 - 2\pi r^3 - 324}{r^2} = \frac{4\pi r^3 - 324}{r^2} = 0; \text{ at optimal point}$$

$$Thus, \quad 4\pi r^3 - 324 = 0; r^3 = \frac{324}{4\pi}; \quad r = \left(\frac{324}{4\pi}\right)^{\frac{1}{3}} = 2.954$$

Example 4

A piece of land along an existing perimeter wall is to be faced using 360 ft long wire. Find the dimensions of the maximum size of the land that can be fenced if the length is parallel to the wall.

Solution

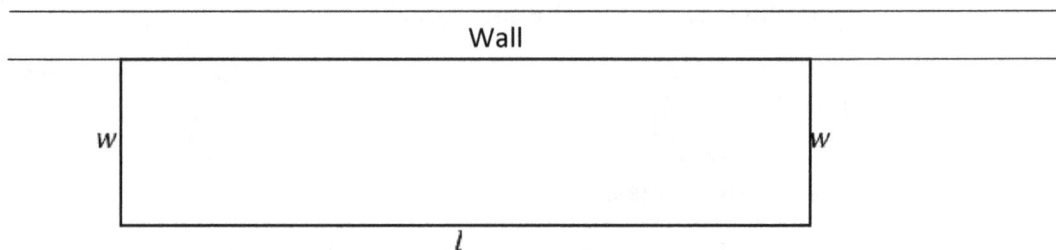

The perimeter to be fenced is $p = l + 2w = 360,$ thus, $\quad l = 360 - 2w$

Area is given by $a = lw$, upon substitution, we get

$$a = w(360 - 2w) = 360w - 2w^2$$

To maximize area, we use derivatives

$$\frac{da}{dw} = 360 - 4w = 0, \quad \text{thus,} \quad 4w = 360, \quad w = 90$$

Since $\frac{d^2a}{dw^2} = -4 < 0,$ the value $w = 90$ is the maximum

Thus, the length is $l = 360 - 2w = 360 - 2(90) = 180 \; ft$

Thus, the dimensions are $180 \; ft \; by \; 90 \; ft$

Section 104.19:The Rolle's Theorem

The theorem sets a condition gives one a guarantee if a function has a minima or a maxima within a given interval. The Rolle 's Theorem states that

If $f(x)$ is a continuous function on a closed interval $[a, b]$ and differentiable on an open interval (a, b) where a and b are distinct real numbers and $f(a)$ and $f(b)$ exists, if $f(a) = f(b)$ where $a < b$, then there is a number k between a and b such that $f'(k) = 0$.

Example 1

Verify Rolle's theorem for $f(x) = \dfrac{2x}{x^2 - 1}$ on $0 \le x \le 2$.

Solution

The function is not continuous on the interval $[0,2]$ since it is not defined at $x = 1$ which is in the interval. Thus, the hypothesis (conditions) of Rolle's is not satisfied. Therefore, Rolle's theorem does not apply.

Example 2

Verify Rolle's theorem for $f(x) = x^2 - 3x + 2$ on $1 \le x \le 2$.

Solution

The function is a polynomial hence it continuous and differentiable on the whole of the real axis including $[1,2]$.

Let $a = 1$ and $b = 2$.

$f(a) = f(1) = (1)^2 - 3(1) + 2 = 0$ and $f(b) = f(2) = (2)^2 - 3(2) + 2 = 0$

Thus, we have $(a) = f(b) = 0$.

The hypothesis of Rolles theorem are satisfied, hence, there is a point c such that $f'(c) = 0$. We find this point.

$f'(x) = 2x - 3$ and $f'(c) = 2c - 3 = 0$, $c = \dfrac{3}{2} = 1.5$

Since 1.5 is in the interval $[1,2]$, $c = 1.5$.

Example 3

Verify Rolle's theorem for $f(x) = x^2 + 2x - \dfrac{21}{4}$ on $\dfrac{3}{2} \le x \le \dfrac{7}{2}$.

Solution

The function is a polynomial hence it continuous and differentiable on the whole of the real axis including $\left[-\dfrac{7}{2}, 3 \right]$.

Let $a = -\dfrac{7}{2}$ and $b = \dfrac{3}{2}$.

$$f(a) = f\left(\dfrac{7}{2}\right) = \left(\dfrac{7}{2}\right)^2 + 2\left(\dfrac{7}{2}\right) - \dfrac{21}{4} = 0 \quad \text{and} \quad f(b) = f\left(\dfrac{3}{2}\right) = \left(\dfrac{3}{2}\right)^2 + 2\left(\dfrac{3}{2}\right) - \dfrac{21}{4} = 0$$

Thus, we have $(a) = f(b) = 0$.

The hypothesis of Rolle's theorem are satisfied, hence, there is a point c such that $f'(c) = 0$. We find this point.

$$f'(x) = 2x + 2; \quad also \; f'(c) = 2x + 2 = 0; \quad c = -1$$

Since -1 is between $-\dfrac{7}{2}$ and $\dfrac{3}{2}$, the required point is $x = -1$.

Example 4

Verify Rolle's theorem for $f(x) = x^2 - x - 6$ on $2 \le x \le 3$.

Solution

The function is a polynomial hence it continuous and differentiable on the whole of the real axis including $[2,3]$.

Let $a = 2$ and $b = 3$.

$$f(a) = f(2) = 2^2 - 2 - 6 = -4 \;, f(b) = f(3) = 3^2 - 3 - 6 = 0 \;.$$

$Since\; f(a) \ne f(b)$, the hypothesis of Rolle's Theorem does not satisfied hence, Rolle's theorem does not apply.

Example 5

Verify Rolle's theorem for $f(x) = x^3 + x^2 - 9x - 9$ on $-3 \le x \le -1$.

Solution

The function is a polynomial hence it continuous and differentiable on the whole of the real axis including $[-3, -1]$.

Let $a = -1$ and $b = -3$.

$f(a) = f(x) = (-1)^3 + (-1)^2 - 9(-1) - 9 = 0$ and
$f(b) = f(-3) = (-3)^3 + (-3)^2 - 9(-3) - 9 = 0$

Thus, $f(a) = f(b)$, hence the Rolle's theorem applies

There is a point c such that $f'(c) = 0$. We determine the point.
$f'(x) = 3x^2 + 2x - 9; \ f'(c) = 3c^2 + 2c - 9$

Using quadratic formula, we have $c = \dfrac{-2 \pm \sqrt{112}}{6} = 1.4305 \ or -2.097$

It is only -2.097 that is in the interval $[-1, -3]$. Thus, the point is $c = -2.097$.

Section 104.20: Mean value theorem

The mean value theorem gives a possibility of having a point on the curve having the tangent that is parallel to the line connecting the two points, which are the endpoints of the interval defined in the theorem. It states that if $f(x)$ is a continuous function on a closed interval $[a, b]$ and differentaiable on an open interval (a, b) for $a < b$, and $f(a)$ and $f(b)$ both exists, then there is a point c in the interval (a, b) such that

$$f'(c) = \frac{f(b) - f(a)}{b - a}$$

We can also modify the results of the theorem then used to estimates roots and other functions. Cross-multiplying, we have $f'(c)(b - a) = f(b) - f(a)$

Thus $f(b) = f(a) + f'(c)(b - a)$

Example 1

Verify the mean value theorem for $f(x) = 2x^2 + 9x + 3$ between $x = 1$ and $x = 4$.

Solution

The function is a polynomial hence it continuous and differentiable on the whole of the real axis including $[1,4]$.

Let $a = 1$ and $b = 4$.

$f(a) = f(1) = 2(1)^2 + 9(1) + 3 = 14$, $f(b) = f(4) = 2(4)^2 + 9(4) + 3 = 71$

Since $f(a)$ and $f(b)$ exists, all the hypothesis of the mean value theorem are satisfied. Therefore, there exists a point, say c , between 1 and 4 such that

$$f'(c) = \frac{f(b) - f(a)}{b - a}$$
$$f'(x) = 4x + 9, \qquad f'(c) = 4c + 9$$

Upon substitution, we have

$$4c + 9 = \frac{71 - 14}{4 - 1} = \frac{57}{3} = 19$$

Thus, $4c = 19 - 9; \quad c = 2.5$

Since 2.5 is between 1 and 4, the value of x being determine is $c = 2.5$.

Example 2

Verify the mean value theorem for $f(x) = e^{2x+1}$ between $x = -0.5$ and $x = 2$.

Solution

The function is an exponential function hence it continuous and differentiable on the whole of the real axis including $[-0.5, 2]$.

Let $a - 0.5$ and $b = 2$.

$f(a) = e^{2(-0.5)+1} = e^0 = 1$ and $f(b) = e^{2(2)+1} = e^5 \approx 148.4$

Thus, $f(a)$ and $f(b)$ both exists hence we can apply the mean value theorem

There is a point, say c, in between -0.5 and 2 such that
$$f'(c) = \frac{f(b) - f(a)}{b - a}$$

$$f'(x) = 2e^{2x+1}, f'(c) = 2e^{2c+1}$$

Upon substitution, we have
$$2e^{2c+1} = \frac{148.4 - 1}{2 - -0.5} = \frac{147.4}{2.5} = 58.96; \quad 2e^{2c+1} = 58.96; \quad e^{2c+1} = 29.48$$

Applying logarithms, we have $\ln 29.48 = 2c + 1; \quad 2c = -1 + \ln 29.48$
$$c = \frac{-1 + \ln 29.48}{2} = 1.192$$

Since 1.192 is between -0.5 and 2, it is the required point.

Example 3

Use the mean value theorem to estimate the roots of

(i). $\sqrt{8}$

(ii). $\sqrt[3]{29}$

Solution

(i). Let the function be $f(x) = \sqrt{x}$, since we want to estimate $\sqrt{8}$, we take it to be $f(b)$ so that $b = 8$.

Then a will be a number less than 8 whose square root is an integers. $f'(c)$ will be approximated using $f'(a)$.

From the above statement, $a = 4$, so that $f(a) = \sqrt{a} = \sqrt{4} = 2$.

However, since 4 is not closer to 8, it is not a good estimate. The best one is 9.

We would like to see what each of the will give us.

$$f(x) = \sqrt{x} = x^{\frac{1}{2}}; \quad f'(x) = \frac{1}{2}x^{-\frac{1}{2}}$$

When $a = 4$

$$f'(a) = f'(2) = \frac{1}{2}\left(4^{-\frac{1}{2}}\right) = \frac{1}{2} \times 2^{2 \times -\frac{1}{2}} = \frac{1}{2} \times 2^{-1} = \frac{1}{2} \times \frac{1}{2} = \frac{1}{4}$$

Using mean value theorem we have $f(b) = f(a) + f'(c)(b - a)$

Upon substitution, we have $f(b) = \sqrt{8} \approx 2 + \frac{1}{4}(8 - 4) = 2 + 1 = 3$

When $a = 9, f(a) = \sqrt{a} = \sqrt{9} = 3$

$$f'(a) = f'(2) = \frac{1}{2}\left(9^{-\frac{1}{2}}\right) = \frac{1}{2} \times \frac{1}{9^{\frac{1}{2}}} = \frac{1}{2} \times \frac{1}{3} = \frac{1}{6}$$

Upon substitution, we have $f(b) = \sqrt{8} \approx 3 + \frac{1}{6}(8 - 9) = 3 - \frac{1}{6} = 2.833$

Which is a bitter approximation that the first one.

Thus $\sqrt{8} \approx 2.833$

(ii). $\sqrt[3]{29}$

Let the function be $f(x) = \sqrt[3]{x}$, since we want to estimate $\sqrt[3]{29}$ we take it to be $f(b)$ so that $b = 29$.

Then a will be a number less than 29 whose cube root is an integer. $f'(c)$ will be approximated using $f'(a)$.

The number 27, $a = 27$ and $f(a) = \sqrt[3]{27} = 3$

$$(x) = \sqrt[3]{x} = x^{\frac{1}{3}}, f'(x) = \frac{1}{3x^{\frac{2}{3}}}; \; f'(c) = f'(27) = \frac{1}{3(27)^{\frac{2}{3}}} = \frac{1}{27}$$

Using mean value theorem we have

$$f(b) = f(a) + f'(c)(b - a) = 3 + \frac{1}{27}(29 - 27) = 3 + \frac{2}{27} = 3.074$$

Thus, $\sqrt[3]{29} \approx 3.074$

Section 104 Conclusion

In this lesson, we have discussed the applications of differentiations, which are optimization and curve sketching. This is done with the help of the concept of turning points and rules of differentiation. We have completed the lessons by looking at the main results in differentiation that describes the highest or lowest points on a curve within a given interval, these are the Rolle's and the mean value theorems.

CALCULUS: APPLICATION OF CONICS AND INTRODUCTION TO DIFFERENTIATION

Content Description

In this session, we will discuss different rules used in differentiation as well the application of differentiation in curve sketching and optimization. We also look at the main theorems in differentiation.

MATH TOPICS

- Calculus 104.16 Rules of differentiation (Reference 1.11).
- Calculus 104.17 Curve Sketching (Reference 1.12).
- Calculus 104.18 Optimization (Reference 1.13).
- Calculus 104.19 The Rolles Theorem (Reference 1.14).
- Calculus 104.20 Mean Value Theorem (Reference 1.15).

INTRODUCTION

There are different types of functions that we meet in both real and academic life; the rate of these functions is referred to as the derivative. Due to different types of function, the basic method of differentiation learnt is not enough to different them in good time, we need some specific viable methods. That is why we need to look at rules of differentiation. As an extension of the concept, we will also look at the application of differentiation as well as some results on differentiation that describes some functions.

Section 104.16: Rules of differentiation

These are rules that enable us carry out differentiation of different types of functions.

Derivative of a constant

The derivative of a $y = f(x) = c$ where c is a constant is $\dfrac{dy}{dx} = 0$

Addition and subtraction

If $y = f(x) \pm h(x)$, then $\dfrac{dy}{dx} = \dfrac{df}{dx} \pm \dfrac{dh}{dx}$

Multiplication by a constant

If $y = kf(x)$ then $\dfrac{dy}{dx} = k\dfrac{df}{dx}$

Product rule

If $y(x) = f(x)g(x)$ then $\dfrac{dy}{dx} = \dfrac{df}{dx}g(x) + \dfrac{dg}{dx}f(x)$

Quotient rule

If $y(x) = \dfrac{f(x)}{g(x)}$ then

$$\dfrac{dy}{dx} = \dfrac{\dfrac{df}{dx}g(x) - \dfrac{dg}{dx}f(x)}{[g(x)]^2}$$

Power rule

If $y(x) = (f(x))^n$ where n is constant, then

$\dfrac{dy}{dx} = n(f(x))^{n-1}\dfrac{df}{dx}$

Chain rule

If $y(x) = f(g(x))$ then

$\dfrac{dy}{dx} = f'(g(x))g'(x) = \dfrac{df}{dg}\dfrac{dg}{dx}$

Exponential differentiation

If $y(x) = e^{u(x)}$, then

$$\frac{dy}{dx} = u'(x)e^{u(x)}$$

If $y(x) = a^{u(x)}$, where a is any other constant other than e then

$$\frac{dy}{dx} = u'(x)e^{u(x)}\log a$$

Differentiation of logarithms

If $y = \ln(u(x))$ then

$$\frac{dy}{dx} = \frac{u'(x)}{u(x)}$$

Trigonometric differentiation

$y = \sin x, y' = \cos x$

$y = \cos x, \qquad y' = -\sin x$

$y = \tan x, y' = \sec^2 x$

$y = \cot x; \quad y' = -\csc x$

$y = \sec x, y' = \sec x \tan x$

$y = \csc x, y' = -\csc x \cot x$

Example 1

Find the derivative of the following functions

(i) $\quad y = \dfrac{2x^2}{4x - 8}$

(ii) $\quad y = (3x + 4)^3(4x - 9)$

(iii) $\quad y = \sin^2(2x + 3)$

(iv) $\quad y = e^{4x + \tan\left(\frac{x}{2}\right)}$

(v) $\quad y = x\ln(3x + 1)$

(vi) $\quad y = \sec\left(e^{(2x-1)^2}\right)$

Solution

(i).We use the quotient rule

$$y = \frac{2x^2}{4x - 8} = \frac{f(x)}{g(x)}$$

Then

$$\frac{dy}{dx} = \frac{\frac{df}{dx}g(x) - \frac{dg}{dx}f(x)}{[g(x)]^2} = \frac{(4x)(4x - 8) - 4(2x^2)}{(4x - 8)^2} = \frac{16x^2 - 32x - 8x^2}{(4x - 8)^2}$$

$$= \frac{8x^2 - 32x}{(4x - 8)^2} = \frac{8x(x - 4)}{(4(x - 2))^2} = \frac{8x(x - 4)}{16(x - 2)^2} = \frac{x(x - 4)}{2(x - 2)^2}$$

Hence $\dfrac{dy}{dx} = \dfrac{x(x - 4)}{2(x - 2)^2}$

(ii) We first use product rule, if $y(x) = f(x)g(x)$ then $\dfrac{dy}{dx} = \dfrac{df}{dx}g(x) + \dfrac{dg}{dx}f(x)$

$$y = (3x + 4)^3(4x - 9)$$

Let $f(x) = (3x + 4)^3$ and $g(x) = 4x - 9$

Using power rule on $(3x + 4)^3, \dfrac{df}{dx} = 3(3x + 4)^2(3) = 9(3x + 4)^2; \quad \dfrac{dg}{dx} = 4$

Thus,

$$\frac{dy}{dx} = \frac{df}{dx}g(x) + \frac{dg}{dx}f(x) = 9(3x + 4)^2(4x - 9) + 4(3x + 4)^3$$

Upon factorization, we get
$$(3x + 4)^2(9(4x - 9) + 4(3x + 4)) = (3x + 4)^2(36x - 81 + 12x + 16)$$

$$= (3x + 4)^2(48x - 97)$$

Therefore, $\dfrac{dy}{dx} = (3x + 4)^2(48x - 97)$

(iii). $y = \sin^2(2x + 3)$

We use chain rule. Let $h = h(x) = 2x + 3$, then $y = \sin^2 h = (\sin h)^2$

Thus, $\dfrac{dh}{dx} = 2$ and by power rule $\dfrac{dy}{dh} = 2\sin h \cos h$

Thus, $\dfrac{dy}{dx} = \dfrac{dy}{dh} \times \dfrac{dh}{dx} = 2(2\sin h \cos h)$

Substituting for h , we get

$$\dfrac{dy}{dx} = 4\sin(2x + 3)\cos(2x + 3)$$

(iv). $y = e^{4x + \tan\left(\frac{x}{2}\right)}$

We use chain rule. Let $h(x) = 4x + \tan\left(\dfrac{x}{2}\right)$, then $y = e^h$

$$\dfrac{dh}{dx} = 4 + \dfrac{d}{dx}\left(\tan\left(\dfrac{1}{2}x\right)\right) = 4 + \dfrac{1}{2}\tan\left(\dfrac{1}{2}x\right)\sec^2\left(\dfrac{1}{2}x\right)$$

$$\dfrac{dy}{dh} = e^h$$

Thus, by chain rule,

$$\dfrac{dy}{dx} = \dfrac{dy}{dh}\dfrac{dh}{dx} = e^h\left(4 + \dfrac{1}{2}\tan\left(\dfrac{1}{2}x\right)\sec^2\left(\dfrac{1}{2}x\right)\right) = e^{4x + \tan\left(\frac{x}{2}\right)}\left(4 + \dfrac{1}{2}\tan\left(\dfrac{1}{2}x\right)\sec^2\left(\dfrac{1}{2}x\right)\right)$$

Hence,

$$\dfrac{dy}{dx} = e^{4x + \tan\left(\frac{x}{2}\right)}\left(4 + \dfrac{1}{2}\tan\left(\dfrac{1}{2}x\right)\sec^2\left(\dfrac{1}{2}x\right)\right)$$

(v). $y = x\ln(3x + 1)$

We use product rule, let $y = f(x)g(x)$ so that $f(x) = x$ and $= g(x) = \ln(3x + 1)$

We use the rule of differentiating functions with natural logarithms

$f'(x) = 1$ and $g'(x) = \dfrac{3}{3x + 1}$

Thus,

$$\dfrac{dy}{dx} = f'(x)g(x) + f(x)g'(x) = 1(\ln(3x + 1)) + \dfrac{3}{3x + 1}(\ln(3x + 1))$$

$$= \ln(3x + 1)\left(1 + \dfrac{3}{3x + 1}\right) = \ln(3x + 1)\left(\dfrac{3x + 1 + 3}{3x + 1}\right) = \ln(3x + 1)\left(\dfrac{3x + 4}{3x + 1}\right)$$

Hence, $\dfrac{dy}{dx} = \left(\dfrac{3x+4}{3x+1}\right)\ln(3x+1)$

(vi). $y = \sec\left(e^{(2x-1)^2}\right)$

We use chain rule

Let $u = e^{(2x-1)^2}$, $y = \sec u$

$\dfrac{du}{dx} = e^{(2x-1)^2}\dfrac{d}{dx}(2x-1)^2 = e^{(2x-1)^2}2(2x-1)(2) = 4(2x-1)e^{(2x-1)^2}$

$\dfrac{dy}{du} = \sec u\tan u$

Thus,

$\dfrac{dy}{dx} = \dfrac{dy}{du} \times \dfrac{du}{dx} = \sec u\tan u\,4(2x-1)e^{(2x-1)^2}$

Upon substitution, we get

$\dfrac{dy}{dx} = \sec\left(e^{(2x-1)^2}\right)\tan\left(e^{(2x-1)^2}\right)4(2x-1)e^{(2x-1)^2}$

Section 104.16: Curve Sketching

Curve sketching is done when the key features of a curve are identified. These features are the intercepts and the turning points.

The intercepts

$x-$ intercept; it is a point where the graph intersects the $x-$ axis. At this point, the $y-$ coordinate is zero.

$y-$ intercept; it is a point where the graph intersects the $y-$ axis. At this point, the $x-$ coordinate is zero.

Turning points

These are points at which the slope of the graph changes. At these points, the slope is zero. These points are maxima, minima and point of inflection.

Maxima

A function $y = f(x)$ has a maxima at $x = a$ is $f'(a) = 0$ and $f''(a) < 0$.

That means around $x = a$, the slope to the left is positive and that to the right is negative.

Minima

A function $y = f(x)$ has a minima at $x = a$ is $f'(a) = 0$ and $f''(a) > 0$.

That means around $x = a$, the slope to the left is negative and that to the right is positive.

Point of inflection

This is a point where the curvature of the curve changes. This may be from concave upwards to downwards or vice verse. If the point of inflection is $x = a$, then $f'(a) = f''(a) = 0$.

To do curve sketching, we have to take not of all these points. We will use examples as illustrations.

Example 1

Sketch the graphs of the following function

(i). $y = 4x^2 - 25$

(ii). $y = 2x^3 + 3x^2 - 9x - 10$

(iii). $y = 2x^3 - 3x^2 - 2x + 3$

Solution

(i). $y = 4x^2 - 25$

$x -$ intercept

At this point, $y = 0$, hence $4x^2 - 25 = 0$; $x^2 = \dfrac{25}{4}, = 6.25$, $x = \sqrt{6.25} = \pm 2.5$ in

$y -$ intercept

At this point, $x = 0$, hence $y = -25$

Maxima or minima

The function has a turning point when $y' = 0$, $y'(x) = 8x = 0$, thus $x = 0$

$y''(x) = 8 > 0$, thus, the turning point is a minima

when $x = 0, y = -25$

The minima is at $(0, -25)$. This also the $y -$ intercept of the function.

Thus, the key features are

$x - intercepts$ at $(2.5, 0)$ and $(-2.5, 0)$, $y -$ intercept and minima both at $(0, -25)$

Skecting the graph based on these, we have

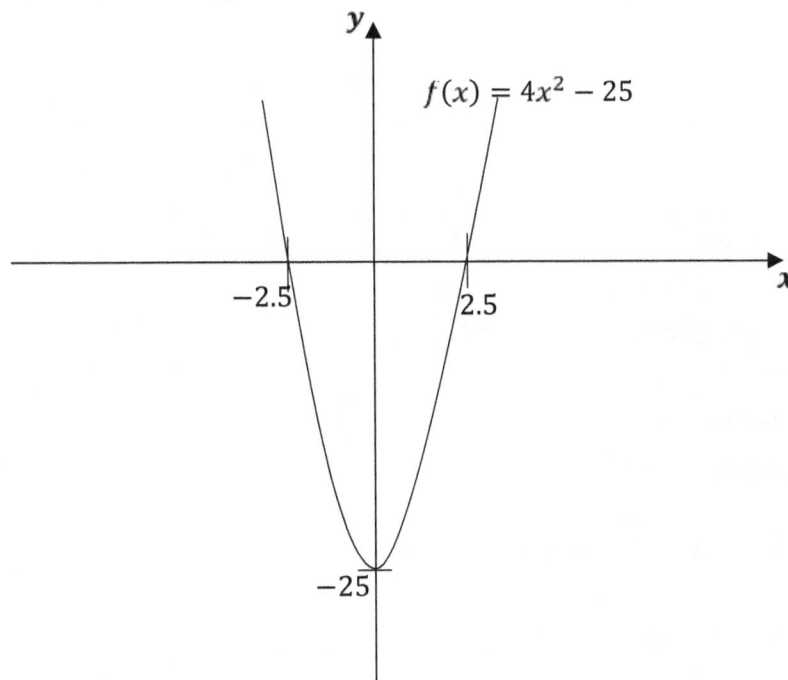

$f(x) = 4x^2 - 25$

(ii). $y = 2x^3 + 3x^2 - 9x - 10$

$y -$ intercept

At this point, $x = 0$, thus $y(0) = 2(0) + 3(0) - 9(0) - 10 = -10$

The point IS $(0, -10)$

$x -$ intercept

At this point, $y = 0$, thus, we solve $2x^3 + 3x^2 - 9x - 10 = 0$

Using rational zero theorem, the roots may be $\pm 1, \pm 2, \mp 5, \mp 10, \mp \dfrac{1}{2}, \mp \dfrac{5}{2}.$

By try 2

$y(2) = 2(2^3) + 3(2^2) - 9(2) - 10 = 0$

Thus, $x = 2$ is a solution implying that $x - 2$ is factor of the polynomial

Using synthetic division, we have

$$
\begin{array}{c|ccc}
2 \quad 2 & 3 & -9 & -10 \\
0 \quad & 4 & 14 & 10 \\
\hline
& 2 \quad 7 & 5 & 0
\end{array}
$$

Thus, when $x - 2$ divides $2x^3 + 3x^2 - 9x - 10$, we get $2x^2 + 7x + 5$

We again factorize the expression

$2x^2 + 7x + 5 = 2x^2 + 5x + 2x + 5 = x(2x + 5) + 1(2x + 5) = (x + 1)(2x + 5)$

Thus, we have $x = -1, \qquad x = -\dfrac{5}{2} = -2.5$

The x intercepts are $x = 2, x = -1$ and $x = -2.5$

Turning points

The function is $y = 2x^3 + 3x^2 - 9x - 10$

$y'(x) = 6x^2 + 6x - 9 = 0$ or $2x^2 + 2x - 3 = 0$

By quadratic formula $x = \dfrac{-2 \pm \sqrt{28}}{4} = \dfrac{-2 \pm 5.292}{4}$

$x = 0.823, x = -1.823$

For $x = 0.823, y = -14.26$

$\quad x = -1.823, y = 4.26$

Second derivative test.

$y''(x) = 12x + 6$

For $x = 0.823, y'' = 15.88 > 0$ Minima

For $x = -1.823, y'' = -15.88 < 0$ Maxima

Point of inflection

$y''(x) = 12x + 6 = 0; x = 0.5$

At $x = 0.5,$ $y = -13.5$

The points are

Intercepts, $(0, -10), (2, 0), (-1, 0)$ and $(-2.5, 0)$

Maxima $(-1.823, 4.26)$

Minima $(0.823, -14.26)$

Point of inflection $(0.5, -13.5)$

The sketch is

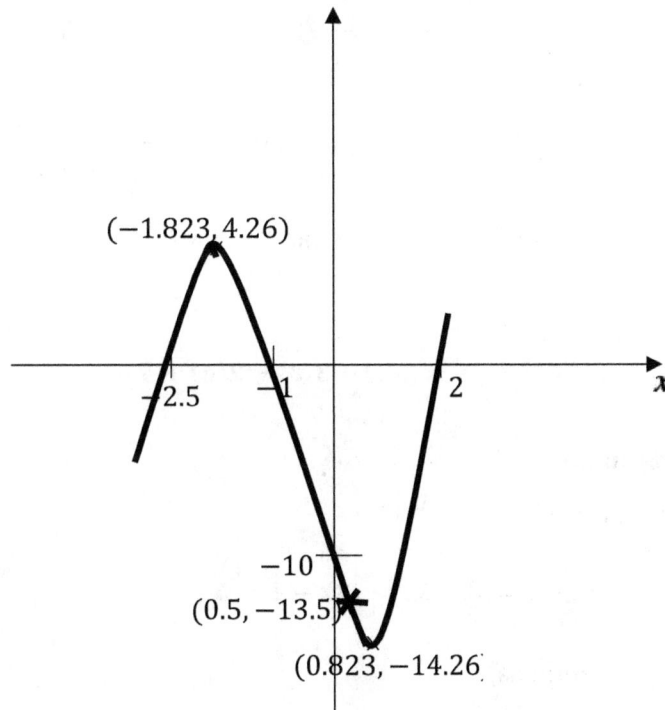

(iii). $y = 2x^3 - 3x^2 - 2x + 3$

Solution

$x-$ intercept

At $x-$ intercept, $y = 0$, hence $2x^3 - 3x^2 - 2x + 3 = 0$. By rational zero theorem, $\pm 1, \pm 3$ or $\pm \dfrac{3}{2}$ is a solution of the polynomial.

$y(1) = 2(1) - 3(1) - 2(1) + 3 = 0$, thus, $x = 1$ is solution implying $x - 1$ is a factor of the polynomial.

Using the direct synthesis method, we get

```
       1 |  2   -3   -2   3
   2  -1  -3 |
       _____
          2   -1   -3   0
```

The quotient is $2x^2 - x - 3$

Factorizing the expression, we get

$2x^2 - x - 3 = 2x^2 - 3x + 2x - 3 = x(2x - 3) + 1(2x - 3) = (x + 1)(2x - 3)$

$Thus, x + 1 = 0 \; and \; 2x - 3 = 0$ implies that $x = -1$ and $x = \dfrac{3}{2} = 1.5$

The $x-$ intercepts are $x = 1, x = -1$ and $x = 1.5$

$y-$ intrercept

This occurs when $x = 0$, $y = 2(0) - 3(0) - 2(0) + 3 = 3$

The turning points

At the turning points $y'(x) = 0$.

$y'(x) = 6x^2 - 6x - 2 = 0; \quad 3x^2 - 3x - 1 = 0$

Using quadratic formula,

$x = \dfrac{3 \pm \sqrt{21}}{6}; \; x = -0.2638; \; x = 1.264$

Second derivative test

$y''(x) = 12x - 6$

$y''(-0.2638) = 12(-0.2638) - 6 < 0$ maxima

$y''(1.264) = 12(1.264) - 6 > 0$ minima

When $x = -0.2638,$ $y = 3.282$

$x = 1.264,$ $y = -0.2821$

Point of inflection

$y''(x) = 12x - 6 = 0;$ $x = 0.5$

When $x = 0.5, y = 1.5$

The points are

Intercepts, $(0,3), (1,0), (-1,0)$ and $(1.5,0)$

Maxima $(-0.2638, 3.282)$

Minima $(1.264, -0.2821)$

Point of inflection $(0.5, 1.5)$

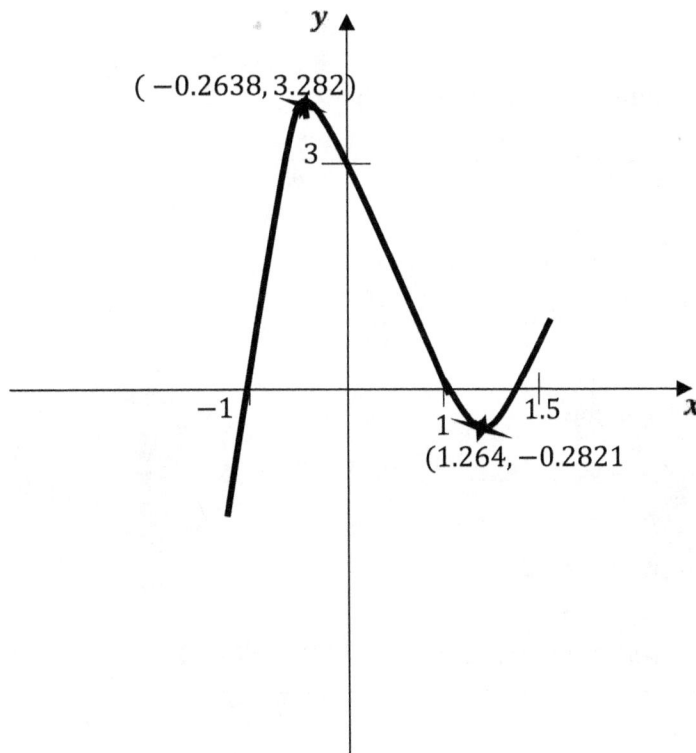

Section 104.18 Optimization

This refers to maximization and minimization of functions. We will use the concepts of maximizing and minimizing functions to optimize variables in the application world.

Some of the most common measures that are optimized include, cost, profit, loss, area, distance or perimeter, volume among others.

We will approach the discussion using examples

Example 1

Minimize the distance around a rectangular piece of land whose area is supposed to be 42 sq. ft

Solution

The distance around the piece of land is perimeter $p = 2(l + w)$

Are is given by $a = lw = 42$; thus $l = \dfrac{42}{w}$

Upon substitution, perimeter would be

$$p(w) = 2\left(\frac{42}{w} + w\right) = \frac{2(42 + w^2)}{w} = \frac{84 + 2w^2}{w}$$

Differentiating with respect to w , we get

$$\frac{dp}{dw} = \frac{4w(w) - (84 + 2w^2)}{w^2} = \frac{2w^2 - 84}{w^2}$$

At the turning point,

$$\frac{dp}{dw} = \frac{2w^2 - 84}{w^2} = 0; \ thus, \qquad 2w^2 - 84 = 0$$

$$w^2 = 42; \quad w = \pm 6.481$$

Since the measure is positive, we take $w = 6.481$

We confirm that this would be the maximum

$$\frac{dp}{dw} = \frac{2w^2 - 84}{w^2}$$

$$\frac{d^2p}{dw^2} = \frac{4w(w^2) - 2w(2w^2 - 84)}{w^4} = \frac{84}{w^4} > 0 \ for \ w > 0$$

Thus, the width perimeter is minimum when $w = 6.381$

The length is $l = \dfrac{42}{w} = \dfrac{42}{6.381} = 6.582\ ft$

Thus, the piece of land measures $6.582\ ft\ by\ 6.481\ ft$

Example 2

A farmer would like to fence her rectangular farm with a 500 feet long wire. Determine the maximum area that she can fence.

Solution

Perimeter is given by $p = 2(l + w)$ where l and w are length and width respectively

Area $a = lw$

We are given that $2(l + w) = 500$, or $l + w = 250$. This implies that $l = 250 - w$

Substituting into the area, we get $a(w) = w(250 - w) = 250w - w^2$

To maximize area, we use the concept of turning points

$\dfrac{da}{dw} = 250 - 2w = 0,$ **at the turing point**

Thus, $2w = 250$; $w = 125ft$

Second derivative test

$\dfrac{d^2a}{dw^2} = -2, thus,$ $w = 125ft$ **is at the maximum**

When $w = 125\ ft,$ $l = 250 - 125 = 125\ ft$

Thus, the maximum area is at $w = 125ft$. From $a = 250w - w^2$, we have

$a_{max} = 250(125) - 125^2 = 15625\ sq. ft$

Example 3

A water tank of capacity 324 cubic feet is to be constructed using the smallest possible amount of material. What would be the radius of this tank?

Solution

The material represents the surface area. Thus, we would like to minimize the surface area (e).

The surface area is given by $s = 2\pi r^2 + 2\pi rh$

We make s a function of one variable, r , hence, we eliminate h

The volume is $v = \pi r^2 h = 324; \quad h = \dfrac{162}{\pi r^2}$

Upon substitution, we have

$$s(r) = 2\pi r^2 + 2\pi r\left(\dfrac{162}{\pi r^2}\right) = 2\pi r^2 + \dfrac{324}{r} = \dfrac{2\pi r^3 + 324}{r}$$

Upon differentiating s , we get

$$\dfrac{ds}{dr} = \dfrac{6\pi r^3 - 2\pi r^3 - 324}{r^2} = \dfrac{4\pi r^3 - 324}{r^2} = 0; \text{ at optimal point}$$

$$Thus, \quad 4\pi r^3 - 324 = 0; r^3 = \dfrac{324}{4\pi}; \quad r = \left(\dfrac{324}{4\pi}\right)^{\frac{1}{3}} = 2.954$$

Example 4

A piece of land along an existing perimeter wall is to be faced using 360 ft long wire. Find the dimensions of the maximum size of the land that can be fenced if the length is parallel to the wall.

Solution

The perimeter to be fenced is $p = l + 2w = 360, \text{thus}, \quad l = 360 - 2w$

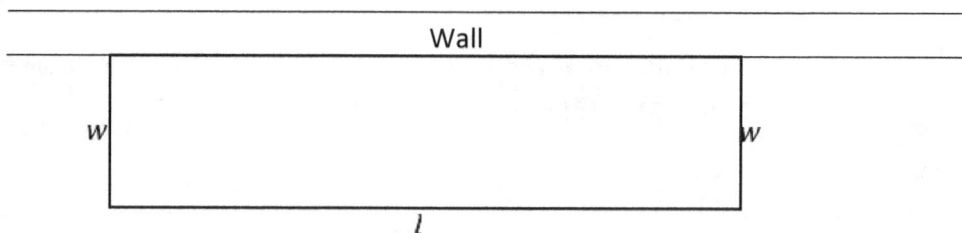

Area is given by $a = lw$, upon substitution, we get

$a = w(360 - 2w) = 360w - 2w^2$

To maximize area, we use derivatives

$$\dfrac{da}{dw} = 360 - 4w = 0, \quad thus, \quad 4w = 360, \quad w = 90$$

Since $\dfrac{d^2a}{dw^2} = -4 < 0,$ the value $w = 90$ is the maximum

Thus, the length is $l = 360 - 2w = 360 - 2(90) = 180\ ft$

Thus, the dimensions are $180\ ft\ by\ 90\ ft$

Section 104.19: The Rolle's Theorem

The theorem sets a condition gives one a guarantee if a function has a minima or a maxima within a given interval. The Rolle 's Theorem states that

If $f(x)$ is a continuous function on a closed interval $[a, b]$ and differentiable on an open interval (a, b) where a and b are distinct real numbers and $f(a)$ and $f(b)$ exists, if $f(a) = f(b)$ where $a < b$, then there is a number k between a and b such that $f'(k) = 0$.

Example 1

Verify Rolle's theorem for $f(x) = \dfrac{2x}{x^2 - 1}$ on $0 \le x \le 2$.

Solution

The function is not continuous on the interval $[0,2]$ since it is not defined at $x = 1$ which is in the interval. Thus, the hypothesis (conditions) of Rolle's is not satisfied. Therefore, Rolle's theorem does not apply.

Example 2

Verify Rolle's theorem for $f(x) = x^2 - 3x + 2$ on $1 \le x \le 2$.

Solution

The function is a polynomial hence it continuous and differentiable on the whole of the real axis including $[1,2]$.

Let $a = 1$ and $b = 2$.

$f(a) = f(1) = (1)^2 - 3(1) + 2 = 0$ and $f(b) = f(2) = (2)^2 - 3(2) + 2 = 0$

Thus, we have $(a) = f(b) = 0$.

The hypothesis of rolles theorem are satisfied, hence, there is a point c such that $f'(c) = 0$. We find this point.

$f'(x) = 2x - 3$ and $f'(c) = 2c - 3 = 0$, $c = \dfrac{3}{2} = 1.5$

Since 1.5 is in the interval $[1,2]$, $c = 1.5$.

Example 3

Verify Rolle's theorem for $f(x) = x^2 + 2x - \dfrac{21}{4}$ on $\dfrac{3}{2} \leq x \leq \dfrac{7}{2}$.

Solution

The function is a polynomial hence it continuous and differentiable on the whole of the real axis including $\left[-\dfrac{7}{2}, 3\right]$.

Let $a = -\dfrac{7}{2}$ and $b = \dfrac{3}{2}$.

$f(a) = f\left(\dfrac{7}{2}\right) = \left(\dfrac{7}{2}\right)^2 + 2\left(\dfrac{7}{2}\right) - \dfrac{21}{4} = 0$ and $f(b) = f\left(\dfrac{3}{2}\right) = \left(\dfrac{3}{2}\right)^2 + 2\left(\dfrac{3}{2}\right) - \dfrac{21}{4} = 0$

Thus, we have $(a) = f(b) = 0$.

The hypothesis of rolles theorem are satisfied, hence, there is a point c such that $f'(c) = 0$. We find this point.

$f'(x) = 2x + 2;\quad also\ f'(c) = 2x + 2 = 0;\quad c = -1$

Since -1 is between $-\dfrac{7}{2}$ and $\dfrac{3}{2}$, the required point is $x = -1$.

Example 4

Verify Rolle's theorem for $f(x) = x^2 - x - 6$ on $2 \leq x \leq 3$.

Solution

The function is a polynomial hence it continuous and differentiable on the whole of the real axis including $[2,3]$.

Let $a = 2$ and $b = 3$.

$f(a) = f(2) = 2^2 - 2 - 6 = -4$, $f(b) = f(3) = 3^2 - 3 - 6 = 0$.

Since $f(a) \neq f(b)$, the hypothesis of Rolle 's Theorem does not satisfied hence, Rolle's theorem does not apply.

Example 5

Verify Rolle's theorem for $f(x) = x^3 + x^2 - 9x - 9$ on $-3 \leq x \leq -1$.

Solution

The function is a polynomial hence it continuous and differentiable on the whole of the real axis including $[-3, -1]$.

Let $a = -1$ and $b = -3$.

$f(a) = f(x) = (-1)^3 + (-1)^2 - 9(-1) - 9 = 0$ and

$f(b) = f(-3) = (-3)^3 + (-3)^2 - 9(-3) - 9 = 0$

Thus, $f(a) = f(b)$, hence the Rolle's theorem applies

There is a point c such that $f'(c) = 0$. We determine the point.

$f'(x) = 3x^2 + 2x - 9; \quad f'(c) = 3c^2 + 2c - 9$

Using quadratic formula, we have $c = \dfrac{-2 \pm \sqrt{112}}{6} = 1.4305 \ or - 2.097$

It is only -2.097 that is in the interval $[-1, -3]$. Thus, the point is $c = -2.097$.

Section 104.20: Mean value theorem

The mean value theorem gives a possibility of having a point on the curve having the tangent that is parallel to the line connecting the two points, which are the endpoints of the interval defined in the theorem. It states that if $f(x)$ is a continuous function on a closed interval $[a, b]$ and differentaiable on an open interval (a, b) for $a < b$, and $f(a)$ and $f(b)$ both exists, then there is a point c in the interval (a, b) such that

$f'(c) = \dfrac{f(b) - f(a)}{b - a}$

We can also modify the results of the theorem then used to estimates roots and other functions. Cross-multiplying, we have $f'(c)(b - a) = f(b) - f(a)$

Thus $f(b) = f(a) + f'(c)(b - a)$

Example 1

Verify the mean value theorem for $f(x) = 2x^2 + 9x + 3$ between $x = 1$ and $x = 4$.

Solution

The function is a polynomial hence it continuous and differentiable on the whole of the real axis including $[1,4]$.

Let $a = 1$ and $b = 4$.

$f(a) = f(1) = 2(1)^2 + 9(1) + 3 = 14$, $f(b) = f(4) = 2(4)^2 + 9(4) + 3 = 71$

Since $f(a)$ and $f(b)$ exists, all the hypothesis of the mean value theorem are satisfied.

Therefore, there exists a point, say c, between $1 \; and \; 4$ such that

$$f'(c) = \frac{f(b) - f(a)}{b - a}$$
$$f'(x) = 4x + 9, \qquad f'(c) = 4c + 9$$

Upon substitution, we have
$$4c + 9 = \frac{71 - 14}{4 - 1} = \frac{57}{3} = 19$$

Thus, $4c = 19 - 9$; $c = 2.5$

Since 2.5 is between 1 and 4, the value of x being determine is $c = 2.5$.

Example 2

Verify the mean value theorem for $f(x) = e^{2x+1}$ between $x = -0.5$ and $x = 2$.

Solution

The function is an exponential function hence it continuous and differentiable on the whole of the real axis including $[-0.5,2]$.

Let $a - 0.5$ and $b = 2$.

$f(a) = e^{2(-0.5)+1} = e^0 = 1$ and $f(b) = e^{2(2)+1} = e^5 \approx 148.4$

Thus, $f(a)$ and $f(b)$ both exists hence we can apply the mean value theorem

There is a point, say c, in between -0.5 and 2 such that
$$f'(c) = \frac{f(b) - f(a)}{b - a}$$

$$f'(x) = 2e^{2x+1}, f'(c) = 2e^{2c+1}$$

Upon substitution, we have

$$2e^{2c+1} = \frac{148.4 - 1}{2 - -0.5} = \frac{147.4}{2.5} = 58.96; \quad 2e^{2c+1} = 58.96; \quad e^{2c+1} = 29.48$$

Applying logarithms, we have $\ln 29.48 = 2c + 1; \quad 2c = -1 + \ln 29.48$

$$c = \frac{-1 + \ln 29.48}{2} = 1.192$$

Since 1.192 is between -0.5 and 2, it is the required point.

Example 3

Use the mean value theorem to estimate the roots of

(i). $\sqrt{8}$

(ii). $\sqrt[3]{29}$

<u>Solution</u>

(i). Let the function be $f(x) = \sqrt{x}$, since we want to estimate $\sqrt{8}$, we take it to be $f(b)$ so that $b = 8$.

Then a will be a number less than 8 whose square root is an integers. $f'(c)$ will be approximated using $f'(a)$.

From the above statement, $a = 4$, so that $f(a) = \sqrt{a} = \sqrt{4} = 2$.

However, since 4 is not closer to 8, it is not a good estimate. The best one is 9.

We would like to see what each of the will give us.

$$f(x) = \sqrt{x} = x^{\frac{1}{2}}; \quad f'(x) = \frac{1}{2}x^{-\frac{1}{2}}$$

When $a = 4$

$$f'(a) = f'(2) = \frac{1}{2}\left(4^{-\frac{1}{2}}\right) = \frac{1}{2} \times 2^{2\times -\frac{1}{2}} = \frac{1}{2} \times 2^{-1} = \frac{1}{2} \times \frac{1}{2} = \frac{1}{4}$$

Using mean value theorem we have $f(b) = f(a) + f'(c)(b - a)$

Upon substitution, we have $f(b) = \sqrt{8} \approx 2 + \frac{1}{4}(8 - 4) = 2 + 1 = 3$

When $a = 9, f(a) = \sqrt{a} = \sqrt{9} = 3$

$$f'(a) = f'(2) = \frac{1}{2}\left(9^{-\frac{1}{2}}\right) = \frac{1}{2} \times \frac{1}{9^{\frac{1}{2}}} = \frac{1}{2} \times \frac{1}{3} = \frac{1}{6}$$

Upon substitution, we have $f(b) = \sqrt{8} \approx 3 + \frac{1}{6}(8-9) = 3 - \frac{1}{6} = 2.833$

Which is a bitter approximation that the first one.

Thus $\sqrt{8} \approx 2.833$

(ii). $\sqrt[3]{29}$

Let the function be $f(x) = \sqrt[3]{x}$, since we want to estimate $\sqrt[3]{29}$ we take it to be $f(b)$ so that $b = 29$.

Then a will be a number less than 29 whose cube root is an integer. $f'(c)$ will be approximated using $f'(a)$.

The number 27, $a = 27$ and $f(a) = \sqrt[3]{27} = 3$

$$(x) = \sqrt[3]{x} = x^{\frac{1}{3}}, f'(x) = \frac{1}{3x^{\frac{2}{3}}}; \; f'(c) = f'(27) = \frac{1}{3(27)^{\frac{2}{3}}} = \frac{1}{27}$$

Using mean value theorem we have

$$f(b) = f(a) + f'(c)(b-a) = 3 + \frac{1}{27}(29-27) = 3 + \frac{2}{27} = 3.074$$

Thus, $\sqrt[3]{29} \approx 3.074$

Section 104 Conclusion

In this lesson, we have discussed the applications of differentiations, which are optimization and curve sketching. This is done with the help of the concept of turning points and rules of differentiation. We have completed the lessons by looking at the main results in differentiation that describes the highest or lowest points on a curve within a given interval, these are the Rolle's and the mean value theorems.

CALCULUS: APPLICATION OF CONICS AND INTRODUCTION TO DIFFERENTIATION

Content Description

In this session, we will discuss the Taylor's polynomial as a prerequisite to series and sequences. This will lead us to their applications in economics. We will also introduce the concepts of ant derivative and look at the methods of integration.

MATH TOPICS

- Calculus 104.21 Taylor's Polynomial (Reference 1.11).
- Calculus 104.22 Common Series and Sequences (Reference 1.12).
- Calculus 104.23 Application of Series and sequences in economics integration (Reference 1.13).
- Calculus 104.24 Ant derivatives (Reference 1.14).
- Calculus 104.25 Methods of integration (Reference 1.15).

INTRODUCTION

We encounter a number of series and sequence that we meet in real-life situation as well as in other fields of study. For instance, amortization and calculation of present values is a series, which can be solved by analytic methods. For us to understand these concepts well, we need to know how to get Taylor's polynomial which also approximate functions around some point. We also need to understand other types of series and sequences as well as their application. Apart from series, we also need to understand the concept of integration, which helps us determine area, volume, intensity among others.

Section 104.21:Taylor's Polynomial

Taylors polynomial is an approximation of any function using a polynomial. Taylor polynomial of degree n is given by $P_n(x) = f(a) + (x-a)f'(a) + \frac{(x-a)^2}{2!}f''(a) + \frac{(x-a)^3}{3!}f'''(a) + \cdots + \frac{(x-a)^n}{n!}f^n(a)$

Given a function $f(x)$, we may have need to estimate its value of a function around a point. This can be done using a polynomial of any degree; it may be a linear function (one degree), a quadratic polynomial (2 degree polynomial), a 3 degree polynomial and so one. The higher the degree of polynomial, the better the approximation.

Let us consider the examples and see how this is done.

Example 1
Find the third and fourth Taylor polynomials for $f(x) = c$ (2) at $x = \pi$.

Solution
To get third and fourth degree polynomials, we differentiate the function three and four times respectively.
$$f(x) = c_i \ (2x)$$
$$f'(x) = -2\sin 2x$$
$$f''(x) = -4\cos(2x)$$
$$f'''(x) = 8\sin(2x)$$
$$f^I\ (x) = 16\cos 2x$$

We evaluate the functions at $x = \pi$
$$f(x) = c \ (2x) = 1$$
$$f'(\pi) = -2\sin 2\pi = 0$$
$$f^{I \cdot (\pi)} = -4\cos(2\pi) = -4$$
$$f''\ ^{(\pi)} = 8\sin(2\pi) = 0$$
$$f^I\ (\pi) = 16\cos 2\pi = 16$$

The third degree Taylor polynomial is
$$P_3(x) = f(a) + (x-a)f'(a) + \frac{(x-a)^2}{2!}f''(a) + \frac{(x-a)^3}{3!}f'''(a)$$

For $a = \pi$, we have
$$P_3(x) = f(\pi) + (x-\pi)f'(\pi) + \frac{(x-\pi)^2}{2!}f''(\pi) + \frac{(x-\pi)^3}{3!}f'''(\pi)$$

Upon substitution, we get

$$P_3(x) = 1 + (x - \pi)(0) + \frac{(x - \pi)^2}{2!}(-4) + \frac{(x - \pi)^3}{3!}(0) = 1 - 2(x - \pi)^2$$

$$= 1 - 2x^2 + 4\pi - 2\pi^2 = -2x^2 + 4\pi + (1 - 2\pi^2)$$

Thus, $P_3(x) = -2x^2 + 4\pi + (1 - 2\pi^2)$

$$P_4(x) = f(\pi) + (x - \pi)f'(\pi) + \frac{(x - \pi)^2}{2!}f''(\pi) + \frac{(x - \pi)^3}{3!}f''(\pi) + \frac{(x - \pi)^4}{4!}f^I(\pi)$$

$$= P_3(x) + \frac{(x - \pi)^4}{4!}f^I(\pi) = -2x^2 + 4\pi + (1 - 2\pi^2) + \frac{(x - \pi)^4}{4!}f^I(\pi)$$

$$P_4(x) = -2x^2 + 4\pi + (1 - 2\pi^2) + \frac{2}{3}(x - \pi)^4$$

Example 2

Find a fourth degree polynomial for $f(x) = e^{-5}$ a $x = 0$.

Solution

To get fourth degree polynomials, we differentiate the function four times.

$$f(x) = e^{-5x}, \qquad f'(x) = -5e^{-5x}; \qquad f''(x) = 25e^{-5x}; f''(x) = -125e^{-5x} f^I(x) = 625e^{-5x}$$

We evaluate the functions at $x = 0$.

$$f(0) = e^{-5(0)} = 1, f'(0) = -5e^{-5(0)} = -5, \quad f''(0) = 25e^{-5(0)} = 25;$$

$$f''(0) = -125e^{-5(0)} = -125 f^I(0) = 625e^{-5(0)} = 625$$

Using the Taylor's polynomial, we have the following polynomial evaluates at $x = 0$.

$$P_4(x) = f(0) + (x - 0)f'(0) + \frac{(x - 0)^2}{2!}f''(0) + \frac{(x - 0)^3}{3!}f''(0) + \frac{(x - 0)^4}{4!}f^I(0)$$

Upon substitution, we get

$$P_4(x) = 1 - 5x + \frac{25}{2}x^2 - \frac{125}{6}x^3 + \frac{625}{24}x^4$$

Example 3

Find a third degree polynomial for $f(x) = x$ x a $x = e$.

Solution

To get third degree polynomials, we differentiate the function three times.

$$f(x) = x \quad x; \quad f'(x) = \ln x + 1 f''(x) = \frac{1}{x} f''(x) = -\frac{1}{x^2}$$

Evaluating at $x = e$, we have

$$f(e) = e \quad e = e; \quad f'(e) = \ln e + 1 = 2 f''(e) = \frac{1}{e} f''(e) = -\frac{1}{e^2}$$

The third degree Taylor polynomial is

$$P_3(x) = f(a) + (x-a)f'(a) + \frac{(x-a)^2}{2!}f''(a) + \frac{(x-a)^3}{3!}f'(a)$$

Upon substitution, we get

$$P_3(x) = e + 2(x-e) + \frac{(x-a)^2 e^{-1}}{2!} - \frac{(x-a)^3 e^2}{3!}$$

Example 4

Find a 3rd degree polynomial of $f(x) = \frac{2}{4x-1}$ at $x = 1$ hence use it to estimate $f(0.4)$

Solution

To get third degree polynomials, we differentiate the function three times.

$$f(x) = \frac{2}{4x-1} = 2(4x-1)^{-1};$$
$$f'(x) = -2(4x-1)^{-2}(4) = -8(4x-1)^{-2}$$
$$f''(x) = 16(4x-1)^{-3}(4) = 64(4x-1)^{-3}$$
$$f'''(x) = 192(4x-1)^{-4}(4) = 768(4x-1)^{-4}$$

Evaluating the functions at $x = 1$, we get

$$f(1) = 2(4(1)-1)^{-1} = 2(3)^{-1} = \frac{2}{3}$$
$$f'(1) = -8(4(1)-1)^{-2} = -\frac{8}{3^2} = -\frac{8}{9}$$
$$f''(1) = 64(4(1)-1)^{-3} = \frac{64}{3^3} = \frac{64}{27}$$
$$f'''(1) = 768(4(1)-1)^{-4} = \frac{768}{3^4} = \frac{768}{81}$$

The third degree Taylor polynomial is

$$P_3(x) = f(a) + (x-a)f'(a) + \frac{(x-a)^2}{2!}f''(a) + \frac{(x-a)^3}{3!}f'''(a)$$

Upon substitution, we get

$$P_3(x) = \frac{2}{3} - \frac{8}{9}(x-1) + \frac{64}{27}\frac{(x-1)^2}{2!} + \frac{768}{81}\frac{(x-1)^3}{3!}$$

$$P_3(x) = \frac{2}{3} - \frac{8}{9}(x-1) + \frac{32}{27}(x-1)^2 + \frac{128}{81}(x-1)^3$$

Section 104.22: Common Series and Sequences

A sequence is a pattern of numbers, letters, or items. Here, we are mostly interested in sequence of numbers. A series is a sum of terms in a sequence.

Thus, the series of the sequence $a_1, a_2, a_3, \ldots$ is $a_1 + a_2 + a_3 + \cdots = \sum_{i-1}^{\infty} a_i$ if the sequence is infinite. It is finite, we have $a_1 + a_2 + a_3 + \cdots + a_n = \sum_{i-1}^{n} a_i$

The most common sequence are geometric and arithmetic sequence where their nth terms are $a_n = ar^{n-1}$ and $a_n = a + (n-1)d$ respectively here a, r and d are the first terms, common ratio and common difference respectively.

Our focus here is on series. The most common series are mostly some Maclaurin series which are Taylor series evaluated at $x = 0$.

These are the geometric series, sine series, cosine series, the series for natural logarithm, the exponential series among others.

Example 1

Generate a Maclaurin series for $y = e^x$ and $y = \dfrac{1}{x-1}$ at $x = 0$.

Solution
The Maclaurin series a regenerated from Taylor series which is "an infinite degree Taylor polynomial." This is in quotations because we cannot have an infinite degree polynomial.

The series is
$$P_\infty(x) = f(a) + (x-a)f'(a) + \frac{(x-a)^2}{2!}f''(a) + \frac{(x-a)^3}{3!}f''(a) + \cdots + \frac{(x-a)^n}{n!}f^n(a)$$
$$+ \frac{(x-a)^{n+1}}{(n+1)!}f^{n+1}(a) + \cdots$$

Taking derivatives, at least 4, we get

$$y = e^x$$

$$y' = y'' = y''' = y^I = e^x$$

Evaluating the functions at $x = 0$, we get $y' = y'' = y'' = y^I = e^0 = 1$
Upon substitution, we get

$$e^x = 1 + x + \frac{x^2}{2!} + \frac{x^3}{3!} + \cdots + \frac{x^n}{n!} + \frac{x^{n+1}}{(n+1)!} + \cdots = \sum_{x=0}^{\infty} \frac{x^n}{n!}$$

$$y = \frac{1}{1-x}$$
$$y = (1-x)^{-1}$$

$$y' = (1-x)^{-2}; \quad y'' = 2(1-x)^{-3}; \quad y''' = 6(1-x)^{-4}; \quad y^I = 24(1-x)^{-5}$$

Evaluating at $x = 0$, we get
$$y(0) = (1-x)^{-1} = 1, y'(0) = (1-0)^{-2} = 1; \quad y''(0) = 2(1-0)^{-3} = 2;$$

$$y''(0) = 6(1-0)^{-4} = 6; \quad y^I(0) = 24(1-x)^{-5} = 24$$

Upon substitution into the formula

$$P_{\infty}(x) = f(a) + (x-a)f'(a) + \frac{(x-a)^2}{2!}f''(a) + \frac{(x-a)^3}{3!}f'''(a) + \cdots + \frac{(x-a)^n}{n!}f^n(a)$$
$$+ \frac{(x-a)^{n+1}}{(n+1)!}f^{n+1}(a) + \cdots$$

We get

$$\frac{1}{1-x} = 1 + x + x^2 + x^3 + x^4 + \cdots + x^n + x^{n+1}\ldots = \sum_{x=0}^{\infty} x^n$$

Below is the some of the common series

$$\sin x = x - \frac{x^3}{3!} + \frac{x^5}{5!} - \cdots = \sum_{x=0}^{\infty} \frac{(-1)^n x^{2n+1}}{(2n+1)!} \text{ valid for all } x$$

$$\cos x = 1 - \frac{x^2}{2!} + \frac{x^4}{4!} - \cdots = \sum_{x=0}^{\infty} \frac{(-1)^n x^{2n}}{(2n)!} \text{ valid for all } x$$

$$e^x = 1 + x + \frac{x^2}{2!} + \frac{x^3}{3!} + \cdots = \sum_{x=1}^{\infty} \frac{x^n}{n!} \text{ valid for all } x$$

$$\frac{1}{1-x} = 1 + x + x^2 + x^3 + \ldots = \sum_{x=0}^{\infty} x^n \text{ valid for} |x| < 1$$

$$\ln(1+x) = x - \frac{x^2}{2} + \frac{x^3}{3} - \cdots = \sum_{x=1}^{\infty} \frac{(-1)^{n+1} x^n}{n} \text{ valid for} |x| < 1$$

Example2

Use the series for $\frac{1}{1-x}$ to find the series for $\frac{2}{1-2}$.

<u>Solution</u>

The series for

$$\frac{1}{1-x} \text{ is } \sum_{x=0}^{\infty} x^n.$$

We need to express $\frac{2}{1-2x}$ in terms of $\frac{1}{1-x}$

$$\frac{2}{1-2x} = 2\frac{1}{1-(2x)} = 2(1 + (2x) + (2x)^2 + (2x)^3 + \ldots = 2\sum_{x=0}^{\infty}(2x)^n \quad v \quad f \quad |2x|$$

$$< 1$$

Since $|2x| < 1$ is equivalent to $|x| < \frac{1}{2} = 0.5$, we have

$$\frac{2}{1-2x} = (1 + (2x) + (2x)^2 + (2x)^3 + \ldots = 2\sum_{x=0}^{\infty}(2x)^n \quad v \quad f \quad |x| < 0.5$$

Example 3

Find the exact value of the series $\frac{x}{3} - \frac{x^3}{1} + \frac{x^5}{3} - \cdots$

<u>Solution</u>

This is an alternating series with the dominator being a multiple of 3. Since we are used to the denominators being in terms of factorial, we can figure out what we should have

$$3 = 3(1!); \quad 18 = 3(6) = 3(3!) \quad a \quad 360 = 3(120) = 3(5!)$$

Thus, we have

$$\frac{x}{3} - \frac{x^3}{18} + \frac{x^5}{360} - \cdots = \frac{1}{3}\left(\frac{x}{1!}\right) - \frac{1}{3}\left(\frac{x^3}{3!}\right) + \frac{1}{3}\left(\frac{x^5}{5!}\right) - \cdots = \frac{1}{3}\left(x - \frac{x^3}{3!} + \frac{x^5}{5!} - \cdots\right)$$

The series inthe bracket is a sine series $\sum_{x=0}^{\infty} \frac{(-1)^n x^{2n+1}}{(2n+1)!}$

Hence we have

$$\frac{1}{3}\left(x - \frac{x^3}{3!} + \frac{x^5}{5!} - \cdots\right) = \frac{1}{3}\sum_{x=0}^{\infty} \frac{(-1)^n x^{2n+1}}{(2n+1)!} = \sum_{x=0}^{\infty} \frac{(-1)^n x^{2n+1}}{3(2n+1)!}$$

$$\sin x = x - \frac{x^3}{3!} + \frac{x^5}{5!} - \cdots = \sum_{x=0}^{\infty} \frac{(-1)^n x^{2n+1}}{(2n+1)!}$$

Therefore,

$$\frac{x}{3} - \frac{x^3}{18} + \frac{x^5}{360} - \cdots = \sum_{x=0}^{\infty} \frac{(-1)^n x^{2n+1}}{3(2n+1)!}$$

Example 4

Estimate the exact value of $1 + 5 + \frac{2}{2} + \frac{1}{6} + \cdots$

<u>Solution</u>

We try to express the series in form of known series

Since $2 = 2!, 6 = 3!, 25 = 5^2$ and $125 = 5^3$

We substitute the values in the series above and see it will be similar to a series that we know

$$1 + 5 + \frac{25}{2} + \frac{125}{6} + \cdots = 1 + 5 + \frac{5^2}{2!} + \frac{5^3}{3!} + \cdots$$

Taking 5 to be x, we have $1 + x + \frac{x^2}{2!} + \frac{x^3}{3!} + \cdots$ which is an exponential series, e^x

Thus $1 + 5 + \frac{5^2}{2!} + \frac{5^3}{3!} + \cdots = e^5 = 148.4$

Section 104.23 Application of Series and sequences in economics integration

Series and sequence are highly applied in economics and finance. We are going to discuss in brief, these applications.

Savings

a). One time saving

When a person makes a saving of amount P at a rate of r, the amount of money at the end of the year would be $a_1 = P(1 + r)$

If the same amount is again saved at the same rate, the amount after one years, that is two years after the initial saving would be

$$a_2 = P(1 + r)(1 + r) = P(1 + r)^2$$

If the same is saved for the third year, we amount at the end of the years would be

$$a_3 = P(1 + r)(1 + r)(1 + 3) = P(1 + r)^3$$

This becomes a geometric series with the first term $P(1 + r)$ and a common ratio of $1 + r$.

Therefore, if P is saved at a compound interest rate of r per each compounding period, then the amount after each compounding period n is

$$a_n = P(1 + r)^n$$

b) Regular interval savings

In this case, a fixed amount of savings is made after each compounding period.

If P is saved after each compounding period at a rate of r per the compounding period, then after n compounding periods (such as a year) the first saving would yield $a_1 = P(1 + r)^n$

Since the second saving take one less year, will yield $a_2 = P(1 + r)^{n-1}$

The third, fouth, second last and last savings would yield $a_3 = P(1 + r)^{n-2}, a_4 = P(1 + r)^{n-3}, a_{n-1} = P(1 + r)^2$ and $a_n = P(1 + r)$

This makes a geometric series

The sum of the savings after n terms is
$$s_n = P(1 + r)^n + P(1 + r)^{n-1} + P(1 + r)^{n-2} + . + P(1 + r)^2 + P(1 + r)$$
Or
$$s_n = P(1 + r) + P(1 + r)^2 + \cdots + P(1 + r)^{n-2} + P(1 + r)^{n-1} + P(1 + r)^n$$

The series has $n + 1$ terms with the first term as $P(1 + r)$ and the common ratio $1 + r$

The series is

$$S_k = \frac{a(r^k - 1)}{r - 1}$$ where a is the first term, r the common ratio and k the period

Upon substitution, we get

$$S_n = \frac{a(r^k - 1)}{r - 1} = \frac{P(1 + r)((1 + r)^k - 1)}{(1 + r) - 1} = \frac{P(1 + r)((1 + r)^k - 1)}{r}$$

$$S_n = \frac{P(1 + r)}{r}((1 + r)^k - 1)$$

Loan repayment

If an amount A is borrowed from a and is to be paid at a rate of r per month (payment period) where d is paid every month, the amount of loan after one moth would be $A(1 + r)$ and the amount remaining after paying d would be

$$a_1 = A(1 + r) - d$$

After the second month, the amount remaining would be

$$a_2 = (A(1 + r) - d)(1 + r) - d = A(1 + r)^2 - d(1 + r) - d$$

After the third month, the amount remaining would be

$$a_3 = (A(1 + r)^2 - d(1 + r) - d)(1 + r) - d = A(1 + r)^3 - d(1 + r)^2 - d(1 + r) - d$$

After n months, the remaining amount would be

$$a_n = A(1 + r)^n - d(1 + r)^{n-1} - d(1 + r)^{n-2} - \cdots - d(1 + r) - d$$

If the loan is fully paid, the remaining amount a_n would be zero if n is the time taken to pay the loan.

Thus, we would have

$$0 = A(1 + r)^n - d(1 + r)^{n-1} - d(1 + r)^{n-2} - \cdots - d(1 + r) - d$$

$$A(1 + r)^n = d(1 + r)^{n-1} + d(1 + r)^{n-2} + \cdots + d(1 + r) + d$$

$$A(1 + r)^n = d + d(1 + r) + \cdots + d(1 + r)^{n-2} + d(1 + r)^{n-1}$$

The right hand side is a geometric series with the first term d, common difference $1 + r$ and n terms.

Using the formula,

$$s_k = \frac{a(r^k - 1)}{r - 1}$$ where a is the first term, r the common ratio and k the period

We have

$$A(1 + r)^n = \frac{d((1 + r)^n - 1)}{(1 + r) - 1} = \frac{d((1 + r)^n - 1)}{r}$$

$$Ar(1 + r)^n = d\left((1 + r)^k - 1\right)$$

$$A = \frac{d((1 + r)^n - 1)}{r(1 + r)^n}$$

$$d = \frac{A\ (1 + r)^n}{(1 + r)^n - 1} \text{ (the monthly payment)}$$

Annuities

The is an asset that pay an amount, whether fixed or variable, at equal intervals of time for a given time, like n. If P is invested at a compound interest of r, then $A = P(1 + r)^n$ is realized after a period, n. The value A is called the future value of P and P, the present value of P.

The sum of the present value of the periodically made payments is the present value of an annuity.

$$P = \frac{A}{1 + r} + \frac{A}{(1 + r)^2} + \frac{A}{(1 + r)^3} + \cdots + \frac{A}{(1 + r)^n}$$

The right hand side a a geometric series with the first terms $\frac{A}{1+r}$, common ratio $\frac{1}{r+1}$ and n terms.

Using summation formula, we have

$$P = \frac{\frac{A}{1+r}\left(1 - \left(\frac{1}{1+r}\right)^n\right)}{1 - \frac{r}{1+r}} = \frac{A\left(1 - \left(\frac{1}{1+r}\right)^n\right)}{r}$$

Thus

$$P = \frac{A\left(1 - \left(\frac{1}{1+r}\right)^n\right)}{r}$$

Example 1

A businessperson invested $2300 for a period of 8 years, how much is he supposed to get if the funds grows at a compound interest of 6% p.a.

Solution

The principal amount invested is $P = \$2300$

Time taken = 8 y

Rate = 6% = 0.06

Amount after 8 years is $A = P(1 + r)^n = 2300(1 + 0.06)^8 = 2300 \times 1.06^8 = \$3665.85$

The amount yield is $3665.85

Example 2

Gentian borrows a loan of $10500. She is supposed to repay it in 20 months. What would be the monthly fixed payments if the loan attracts an interest of 15% per year.

Solution

Monthly payments are given by

$$d = \frac{A\ (1 + r)^n}{(1 + r)^n - 1}$$

Where, A is the loan amount, r the interest rate and n the repayment period

We have $A = \$10500, n = 20\ m\quad hs$

$$r = \frac{15}{100} p.a = \frac{15}{12 \times 100} p \quad m \quad h = \frac{5}{400} = 0.0125$$

Upon substitution, we get

$$d = \frac{10500 \times 0.0125(1 + 0.0125)^2}{(1 + 0.0125)^2\ - 1} = \$596.6$$

The monthly payments are $596.6

Example 3

Judies is expected to pay a total of $1960 for her son's school fees in ten years to come. However, she does not have all the money to pay when that times comes. He decides to save funds that would earn her 7.5% compound interest for the 10 years. How much is she supposed to pay save now to get the exact amount?

Solution
The expected amount to pay is the future value   A = $1960

The rate at which the savings will grow is 7.5% = 0.075

Time taken for the funds to multiply is 10 years

Let the savings be p

Then

$$P = \frac{A}{(1+r)^n} = \frac{1960}{(1+0.075)^1} = 951.0$$

The savings would be $951

Example 4
Alison makes a month payment of $385 for her investment insurance plan that earns her 1.25% every month. How much would her funds accumulate to after 20 years?

Solution
Each month, $385 = d is paid. This is a regular type of investment

The interest is $r = 0.0125$

Time taken is $n = 20 \times 12 = 240$ months

The amount that accumulates after the given time is given by

$$A = \frac{d((1+r)^n - 1)}{r(1+r)^n}$$

Upon substitution, we have

$$A = \frac{385((1+0.0125)^2 \quad -1)}{r(1+r)^2} = \$29237.78$$

Section 104.24: Antiderivatives

Antiderivative is a term used to imply the reverse process of determining a derivative. Since the derivative determines the rate of change (slope), with antiderivative, assumes that one has the rate of change and wishes to determine the original function.

Remembers that the derivative of a constant function $f(x) = c$ is zero. Since antiderivative must take us back to the initial function, we must have add c to the new function, where c takes care of any constant the function had. Therefore, all aintiderivatives must have a constant c added to it. The other name of antiderivative is indefinite integral.

If we have a function $y' = ax^n$, then its antiderivative would be $y = \frac{ax^{n+1}}{n+1} + c$

We have

$$\frac{d}{d} = ax^n$$

Multiply through by d t. g $\quad d = ax^n d$

Finding the antiderivative of both sides, we denote this by an integral sign

$$\int d = \int ax^n d$$

The answer would be

$$y = \frac{a}{n+1} x^{n+1} + c$$

The following are the antiderivatives of trigonometric functions

$$\int \cos x \, d = \sin x + c$$

$$\int \sin x \, d = -\cos x + c$$

$$\int \sec^2 x \, d = \tan x + c$$

$$\int \csc^2 x \, d = -\cot x + c$$

$$\int \sec x \tan x \, d = \sec x + c$$

$$\int \csc x \cot x \, d = -\csc x + c$$

Like integrals, the following rules hold

If $\int g(x)d = h(x) + c$, $\int f(x)d = p(x) + c$ and k and t are constants

(i). $\int g(x) \pm f(x) d = h(x) + p(x) + t$,

(ii). $\int k \ (x) = kh(x) + t$,

Example
Find the antiderivatives of the following

(i). $f(x) = 2x + 3$

(ii). $y = 5x^4 - 6x + 2$

(iii). $y = \frac{3}{x^2} + 6x^2 + x - 7$

(iv). $y = 2 \ s \ c^2 \ x$

(v). $y = \frac{5s \ x \ t \iota \ x}{6}$

<u>Solution</u>

We apply the method $y = \frac{ax^{n+1}}{n+1} + c$ to functions in (i) up to (iii).

(i). $f(x) = 2x + 3$

The antiderivative is

$$y = \int (2x + 3) \ d = \frac{2x^2}{2} + 3x + c = x^2 + 3x + c$$

(ii). $y = 5x^4 - 6x + 2$

The antiderivative is

$$g = \int (5x^4 - 6x + 2) \ d = \frac{5x^5}{5} - \frac{6x^2}{2} + 2x + c = x^5 - 3x^2 + 2x + c$$

(iii). $y = \frac{3}{x^2} + 6x^2 + x - 7$

The antiderivative is

$$g = \int \left(\frac{3}{x^2} + 6x^2 + x - 7\right) d = \int (3x^{-2} + 6x^2 + x - 7) \ d$$

$$= \frac{3x^{-2+1}}{-1} + \frac{6}{3}x^3 + \frac{2}{2}x^2 - 7x + c = -3x^{-1} + 2x^3 + x^2 - 7x + c$$

The antiderivative is $g = -\frac{3}{x} + 2x^3 + x^2 - 7x + c$

(iv). $y = 20s\ c^2\ x$

The antiderivative is

$$g = \int 20\sec^2 x\ d\ = 20\int \sec^2 x\ d\ = 20\tan x + c$$

(v). $y = \dfrac{5\ s_{\iota}\ \ x\ t_{\iota}\ \ x}{6}$

The antiderivative is

$$g = \int \frac{5\ s_{\iota}\ \ x\ t_{\iota}\ \ x}{6}\ d\ = \int \frac{5}{6}\ s_{\iota}\ \ x\ t_{\iota}\ \ x\ d\ = \frac{5}{6}\sec x + c$$

Section 104.25: Methods of integration

Methods of integration. This refers to the techniques used to integrate functions. The techniques value depending on the nature of the function. We will discuss a few.

Logarithmic functions and exponential functions

$$\int e^a\ d\ = \frac{1}{a}e^a\ + c$$
$$\int \frac{1}{x}d\ = \ln x + c$$

Substitution

This refers of a change of a variable using suitably defined functions. The new variable is usually the expression with the highest power, in roots or the power of an exponential function where applicable. This is illustrated using examples.

Example 1
Integrate the following

(i). $\int s\ (\pi\ )d$

(ii). $\int 9x(3x-1)^2(3x-1)^3\ d$

(ii). $\int x^{\ 4x^2+1}\ d$

Solution

We take the expression with the highest power to be our new variable.

(i). let $u = \pi$, $d = \pi$; $\dfrac{d}{\pi} = d$

Upon substitution, we get

$$\int \sin(\pi\ )\,d\ = \int \sin u \dfrac{d}{\pi} = \dfrac{1}{\pi}\int \sin u\,d\ = -\dfrac{1}{\pi}\cos u + c$$

Upon substituting back, we get

$$\int \sin(\pi\ )\,d\ = -\dfrac{\cos(\pi\ )}{\pi} + c$$

(ii). $\displaystyle\int 9x(3x^2 - 1)^2(3x^2 - 1)^3\,d\ = \int 9x(3x^2 - 1)^5\,d\ = \int 9(3x^2 - 1)^5\,x\,d$

We take the expression with the highest power to be our new variable.

Let $u = 3x^2 - 1\,d\ = 6x\,dx$; $\dfrac{d}{6} = x\,d$

We then substitute for $x\,d$ to get

$$\int 9u^5 \dfrac{d}{6} = \dfrac{3}{2}\int u^5\,d\ = \dfrac{3}{2} \times \dfrac{u^6}{6} + c = \dfrac{u^6}{4} + c$$

Substituting back for u, we get

$$\int 9x(3x^2 - 1)^2(3x^2 - 1)^3\,d\ = \dfrac{3x^2 - 1}{4} + c$$

(ii). $\displaystyle\int x\ ^{4x^2+1}\,d$

The expression with the highest power is the power of e, hence $u = 4x^2 + 1$; $d\ = 8x\,d$ or $\dfrac{d}{8} = x\,d$

Substituting back, we get

$$\int x\ ^{4x^2+1}\,d\ = \int e^{4x^2+1}\,x\,d\ = \int e^u \dfrac{d}{8} = \dfrac{1}{8}\int e^u\,d\ = \dfrac{1}{8}e^u + c$$

Substituting back, we get

$$\int x\ ^{4x^2+1}\,d\ = \dfrac{1}{8}e^{4x^2+1} + c$$

Integration by parts

If the integral can be expressed as a product of u and d , then we use the formula
$$\int u = u - \int v$$

The variable u is chosen so that d is simpler than u.

Example 2
Simplify
$$\int 2x^2 l\iota\ x d$$

Let $u = \ln x$ then $d = \frac{1}{x}d$

Then $d = 2x^2$ and $v = \int 2x^2 d = \frac{2}{3}x^3$

Using
$$\int u = u - \int v \quad \text{, we have}$$
$$\int 2x^2 l\iota\ x d = \frac{2}{3}x^3 \ln x - \int \frac{2}{3}x^3 \times \frac{1}{x}d = \frac{2}{3}x^3 \ln x - \int \frac{2}{3}x^2 d$$
$$= \frac{2}{3}x^3 \ln x - \frac{2}{3}\int x^2 d$$
$$= \frac{2}{3}x^3 \ln x - \frac{2}{3} \times \frac{x^3}{3} + c$$
$$= \frac{2}{3}x^3 \ln x - \frac{2x^3}{9} + c$$

Using trigonometric identities

Trigonometric identities are can also be used to solve trigonometric integrals. Let us consider the example below.

Example 3
Intergrate $\int c\ ^2 x s\ ^2 x d$

Solution
We express all functions in terms of cosine suing the identity
$$\sin^2 x = \frac{1}{2} - \frac{1}{2}\cos 2x, \text{and } \cos^2 x = \frac{1}{2} + \frac{1}{2}\cos 2x$$

$$\int \left(\frac{1}{2} + \frac{1}{2}\cos 2x\right)\left(\frac{1}{2} - \frac{1}{2}\cos 2x\right)d = \int \left(\frac{1}{4} - \frac{1}{4}\cos^2 2x\right)d$$

Since $\cos^2 x = \frac{1}{2} + \frac{1}{2}\cos 2x$; $\cos^2 2x = \frac{1}{2} + \frac{1}{2}\cos 4x$, upon substitution, we get

$$\int \left(\frac{1}{4} - \frac{1}{4}\left(\frac{1}{2} + \frac{1}{2}\cos 4x\right)\right) d = \int \left(\frac{1}{4} - \frac{1}{8} - \frac{1}{8}\cos 4x\right) d = \int \left(\frac{1}{8} - \frac{1}{8}\cos 4x\right) d$$

$$= \frac{1}{8}x - \frac{1}{32}\sin 4x + c$$

Thus, $\int \cos^2 x \sin^2 x\, d = \frac{1}{8}x - \frac{1}{3}\sin 4x + c$

Partial fraction decomposition

This is used to solve integrals where the integrand is a fraction and the denominator has factors. The integration is then done for the individual fractions.

Example 4
Solve

$$\int \frac{2}{x^2 - 3x} d$$

Solution

$$x^2 - 3x = x(x - 3)$$

The fraction can be expressed as a sum of two fractions where the denominators are the factors x and $x - 3$ with denominators to be determined. Since the powers of the denominator is 1. The numerator must have a power less than that of denominator by 1, that is constants. Let the constants be A and B. Thus, we have

$$\frac{2}{x^2 - 3x} = \frac{A}{x} + \frac{B}{x - 3} = \frac{A(x - 3) + B}{x(x - 3)}$$

Thus, we must have $2 = A(x - 3) + B$

We select values of x suitably to eliminate one constant at a time

When $x = 3$, we have $2 = 3B$, $B = \frac{2}{3}$

When $x = 0$, we have $2 = -3A$, $A = -\frac{2}{3}$

Thus, we have

$$\frac{2}{x^2 - 3x} = -\frac{2}{3} \times \frac{1}{x} + \frac{2}{3} \times \frac{1}{x - 3}$$

Thus,

$$\int \frac{2}{x^2 - 3x} d = -\frac{2}{3}\int \frac{1}{x} d + \frac{2}{3}\int \frac{1}{x-3} d = -\frac{2}{3}ln\ x + \frac{2}{3}ln\ (x-3) + c$$

$$= \frac{2}{3}ln\ \frac{1}{x} + \frac{2}{3}ln\ (x-3) + c = \frac{2}{3}ln\frac{x-3}{x} + c = ln\left(\frac{x-3}{x}\right)^{\frac{2}{3}} + c$$

Inverse trigonometric functions

This are used to solve problems such as $\int \frac{1}{u^2+a^2} d$, $\int \frac{1}{u\sqrt{u^2-a^2}} d$ and, $\int \frac{1}{\sqrt{a^2-u^2}} d$

When $\int \frac{1}{u^2+a^2} d$, we take $u = \tan x$ then carry out integration. We expect to get

$$\frac{1}{a}\tan^{-1}\left(\frac{u}{a}\right) + c$$

When $\int \frac{1}{u\sqrt{u^2-a^2}} d$, we take $u = \sec x$ then carry out integration. We expect to get

$$\frac{1}{a}s\ c^{-1}\left(\frac{u}{a}\right) + c$$

When $\int \frac{1}{\sqrt{a^2-u^2}} d$, we take $u = \sin x$ then carry out integration. We expect to get

$$s_i \ ^{-1}\left(\frac{u}{a}\right) + c$$

The function is a polynomial hence it continuous and differentiable on the whole of the real axis including$[1,4]$.

Section 104 Conclusion

In this lesson, we have discussed the Taylor series that opened doors to other common series that we have also discussed. We then moved further to look at the applications of these series, especially, the geometric series in economics. We have finalized by looking at integrations, which is a reverse of differentiation.

CALCULUS: DEFINITE INTEGRALS AND APPLICATION

Content Description

In this session, we will discuss the Definite integrals and their application in business and economic world. We would then complete the lessons by highlighting a few concepts of algebra and set theory used in calculus.

MATH TOPICS

- Calculus 104.26 Definite integration (Reference 1.11).
- Calculus 104.27 Application of definite integrals in the business world (Reference 1.12).
- Calculus 104.28 Application to Economics (Reference 1.13).
- Calculus 104.29 Appendix items, Basic Set Theory, Algebraic Rules (Reference 1.14).

INTRODUCTION

There is no doubt that determination of area , flow of material across the membranes, volume among others things is one of the most common things that advanced graphic designers and engineers do using the concept of calculus. This is an application of definite integral. Apart from this, there is also analysis of areas under curves that is highly applied in the business world. We would like to look at a few of these cases as we explore the application of definite integrals.

Section 104.21: Definite integration

A definite integral is an integral problem having the limits within which the integral is to be solved. When the limits are numbers, then the integral gives a numerical number. It is generally represented as

$$\int_{x=a}^{x=b} f(x)d$$

where $x = a$ is called the lover limit, $x = b$ the upper limit and $f(x)$ the integrand

If $f(x)$ is the a curve of a function, then the definite integral represents the area under the graph of $f(x)$, above $y = 0$ and bounded by lines $x = a$ and $x = b$.

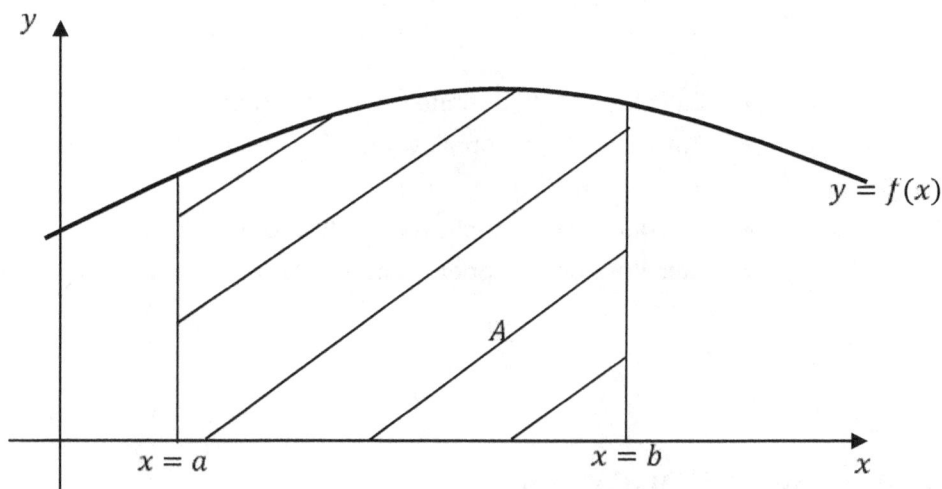

Area of A is given by

$$A = \int_{x=a}^{b} f(x)d$$

Example 1

Evaluate the integrals

(i).

$$\int_1^3 x^2\sqrt{4x^3+1}\,d$$

(ii).

$$\int_0^{\frac{\pi}{2}} \sin 2x\,d$$

(iii).

$$\int_0^{\frac{\pi}{8}} \sec^4 2x \tan 2x\,d$$

(iv).

$$\int_0^1 \frac{1}{3+27x^2}\,d$$

(iv).

$$\int_0^{\pi} x^2 \sin x\,d$$

Solution

(i). $\int_1^3 x^2\sqrt{4x^3+1}\,d$

We integrate by substitution by letting $u = 4x^3 + 1$.

$$\frac{d}{d} = 12x^2$$

Then $x^2 d \ = \frac{1}{1}d$.

Since $4x^3 + 1 = u, u = 5$ when $x = 1$ and 109 when $x = 3$.

Substituting these values we have,

$$\int_5^1 u^{\frac{1}{2}}(\frac{1}{12}d \ )$$

This can be rewritten as,

$$\frac{1}{12}\int_5^1 u^{\frac{1}{2}}.\,d$$

$$= \frac{2}{3}\times\frac{1}{12}\left[u^{\frac{3}{2}}\right]_5^1$$

$$= \frac{1}{18}(109^{\frac{3}{2}} - 5^{\frac{3}{2}})$$

$$= 62.60$$

(ii). $\int_0^{\frac{\pi}{2}}\sin 2x\,d$

$$\int_0^{\frac{\pi}{2}}\sin 2x\,d = \left[-\frac{1}{2}\cos 2x\right]_0^{\frac{\pi}{2}}$$

$$= \left(-\frac{1}{2}\cos\pi\right) - \left(-\frac{1}{2}\cos 0\right)$$

$$= \frac{1}{2} - \left(-\frac{1}{2}\right) = 1$$

(iii).

$$\int_0^{\frac{\pi}{8}}\sec^4 2x\tan 2x\,d$$

We will use the identity $\sec^2 2x = 1 + \tan^2 2x$

$$\int_0^{\frac{\pi}{8}}\sec^4 2x\tan 2x\,d = \int_0^{\frac{\pi}{8}}\tan 2x\sec^2 2x\sec^2 2x\,d$$

$$= \int_0^{\frac{\pi}{8}}\tan 2x\,(1 + \tan^2 2x)\sec^2 2x\,d$$

Let $u = \tan 2x$ so that $d = 2\sec^2 2x\,d$

$$\int_0^{\frac{\pi}{8}}\tan 2x\,(1 + \tan^2 2x)\sec^2 2x\,d = \frac{1}{2}\int_0^1 u(1 + u)\,d = \frac{1}{2}\int_0^{\pi}(u + u^2)\,d$$

The limits change since when $x = 0, u = \tan 2(0) = 0$ and when $x = \frac{\pi}{8}, u = \tan\left(\frac{\pi}{4}\right) = 1$

$$= \frac{1}{2}\left[\frac{1}{2}u^2 + \frac{1}{3}u^3\right]_0^1$$

$$= \frac{1}{2}\left(\frac{1}{2}(1)^2 + \frac{1}{3}(1)^3\right) - 0$$

$$= 2\left(\frac{5}{6}\right) - 0 = \frac{5}{3}$$

(iv).

$$\int_0^1 \frac{1}{3 + 27x^2}\, d$$

We will use the identity

$$\int \frac{1}{a^2 + x^2}\, d = \frac{1}{a}\tan^{-1}\left(\frac{x}{a}\right) + C$$

To make the coefficient of x^2 1 we factor out $\frac{1}{2}$ to get,

$$\int_0^1 \frac{1}{3 + 27x^2}\, d = \frac{1}{27}\int_0^1 \frac{1}{\frac{1}{9} + x^2}\, d$$

$a^2 = \frac{1}{9}$ Thus $a = \frac{1}{3}$

$$\frac{1}{27}\int_0^1 \frac{1}{\frac{1}{9} + x^2}\, d = \left(\frac{1}{27} \times 3 \tan^{-1} 3x\right)_0^1$$

$$= \left(\frac{1}{9}\tan^{-1} 1\right) - \left(\frac{1}{9}\tan^{-1} 0\right)$$

$$= \frac{1}{9} \times \frac{\pi}{4} = \frac{\pi}{36}$$

(v).

$$\int_0^\pi x^2 \sin x \, d$$

We will use integration by parts by letting x^2 be u and $\sin x$ be d

$$\int u \quad = u \quad - \int v$$

$v = -\cos x, d = 2x$

$$\int_0^\pi x^2 \sin x \, d = [-x^2 \cos x]_0^{\frac{3\pi}{4}} + \int_0^\pi 2x \cos x \, d$$

Let $u = 2x$ and $d = \cos x$ then $d = 2d$, $v = \sin x$

We further integrate the integrand on the right hand side.

$$\int_0^\pi x^2 \sin x \, d = [-x^2 \cos x]_0^\pi + [2x \sin x]_0^\pi - \int_0^\pi 2 \sin x \, d$$

$$\int_0^\pi x^2 \sin x \, d = [-x^2 \cos x]_0^\pi + [2x \sin x]_0^\pi + [2 \cos x]_0^\pi$$

$$\int_0^\pi x^2 \sin x \, d = [-x^2 \cos x + 2x \sin x + 2 \cos x]_0^\pi$$

$$= (-(\pi)^2 \cos(\pi) + 2(\pi) \sin(\pi) + 2 \cos(\pi)) - (-(0)^2 \cos 0 + 2(0) \sin 0 + 2 \cos 0)$$

$$= \pi^2 - 2 - 2$$

$$= \pi^2 - 4$$

Calculus 104.27: Application of definite integrals in the business world

Definite integrals have many applications in business world. We will discuss some of these applications using examples.

Let us first discuss some of the terms used in business.

Annuity

This discrete amount of money deposited to an account at equal intervals of time for a given period of time earning a certain interest within the period.

Future value of annuity

This is the total money deposited plus the interest that has accumulated over a certain period.

Example 1

Carrie transfers money continuously into her bank account at the rate of 1,500 dollars per year. If the account earns interest rate of **7%** per annum compounded continuously, calculate the amount of money in Carrie's account after three years.

Solution

If S dollars earns 7% interest compounded continuously, the amount of money after t years will be $Se^{0.0\ t}$

We divide the three years interval $0 \le t \le 3$ into n equal sub intervals each of length $\Delta_{n}t$ years.

If t_i denotes the start of every sub interval, then in each sub interval the money deposited would grow to $(1500\Delta_{n}t)e^{0.0\ (3-t_i)}$.

The future value of the annuity would be equivalent to the sum of money in all sub intervals which is equivalent to $\int_0^3 1500e^{0.0\ (3-t)}d$

$$\int_0^3 1500e^{0.0\ (3-t)}\,d = \int_0^3 1500e^{(0.2\ -0.0\ t)}\,dt$$

Let $u = 0.21 - 0.07t$, $d = \dfrac{d}{-0.0}$

$$\int_0^3 1500e^{(0.2\ -0.0\ t)}\,d = -\frac{1500}{0.07}\int_{0.2}^0 e^u\,d$$

The limits change since when $x = 0, u = 0.21 - 0.07(0) = 0.21$ and when $x = 3, u = 0$

$$-\frac{1500}{0.07}\int_{0.2}^{0} e^u \, d = -\frac{1500}{0.07}[e^u]_{0.2}^{0}$$

$$= -\frac{10}{0.0}(e^0 - e^{0.2}) = -\frac{1}{0.0}(1 - 1.2337) = 5{,}007.40 \text{ Dollars}$$

Example2

A man has a supermarket and a hotel. The supermarket generates the profit at a rate of $P(t) = 2 + t^2$ while the hotel generates the profit at a rate of $P(t) = 1 + 2t$ where t is time in months. Calculate the profit that the supermarket will have generated by the time its profit exceeds the profit from the hotel.

Solution

When the profit from the supermarket will exceeds the profit from the hotel the two profit functions will be equal.

Therefore, $100 + 2t = 20 + t^2$

$$t^2 - 2t - 80 = 0$$

Solving this equation we get $t = 10$ and $t = -8$

We will use $t = 10$ months since it is impossible to have -8 months.

At the interval $0 \leq t \leq 10$ the profit generated by the supermarket will be

$$\int_0^1 20t + t^2 d = \left[20 + \frac{1}{3}t^3\right]_0^1$$

$$= \left(200 + \frac{1000}{3}\right) - 0$$

533.30 Dollars

Example 3

A businessman determines that t weeks after a sales promotion, the sales were
$S(t) = t^2\sqrt{3t^3}$. Find the sales in the first 5 weeks after the sales promotion.

Solution

We find $\int_0^5 t^2\sqrt{3t^3}$

We will use integration by substitution by letting $t^2 = \frac{1}{9}d$ and $u = 3t^3$

$$\int_0^5 t^2\sqrt{3t^3} = \frac{1}{9}\int_0^3 u^{\frac{1}{2}}d = \frac{2}{27}\left[(u)^{\frac{3}{2}}\right]_0^3$$

The limits change since when $t = 0, u = 3(0)^3 = 0$ and when $x = 3, u = 3(5)^3 = 375$

$$= \frac{2}{27}\left[(u)^{\frac{3}{2}}\right]_0^3 = \frac{2}{27}(375)^{\frac{3}{2}} - \frac{2}{27}(0)^3)^{\frac{3}{2}}$$

$$= 5 \quad .9 \quad \text{Dollars}$$

Example 4

A man has a commuter services business. A vehicle generates profit at a rate $P'(t) = 4 \quad -$
$3\ t^2$ per annum. The operating costs relating to the vehicle accumulates at a rate of
$C'(t) = 8 \quad +2\ t^2$ per annum. Find the net profit that the vehicle will have accumulated
during its useful period.

Solution

After its useful life the rate of operating costs will exceed the rate of profit.

At the point where the operating profits exceeds the profit, $P'(t) = C'(t)$

$$4000 - 30t^2 = 800 + 20t^2$$
$$50t^2 = 3200$$
$$t = 8 \text{ years}$$

Thus the useful life of the vehicle is 8 years.

We can sketch the two curves as below

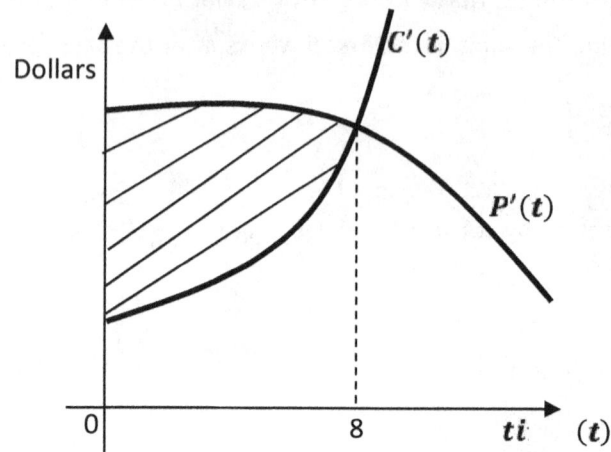

The shaded part represents the net profit during the vehicle's useful period.

Net profit $= \int_0^8 \big(P'(t) - C'(t)\big)d$

$$= \int_0^8 (4000 - 30t^2 - (800 + 20t^2))d$$

$$= \int_0^8 (3200 - 50t^2)d$$

$$= \left[3200t - \frac{50}{3}t^3\right]_0^8$$

$$= (1,7066.70 - 0) = 1,7066.70 \text{ Dollars}$$

Calculus 104.28 Application to Economics

Definite integrals are highly applied in economics. We are going to discuss in brief, some of these applications.

Demand and supply

The demand of a commodity is the quantity of the commodity that consumers are willing and able to buy at certain price.

The supply of a commodity is the quantity of the commodity that the suppliers can supply at a certain price.

When at a certain price the quantity of the commodity that the consumers are willing and able to buy is equal to the quantity that the suppliers can supply, the demand and supply are said to be at equilibrium.

If a commodity that is demanded at a price of 3 Dollars is supplied at a price of 5 Dollars, the extra 2 Dollars is called the suppliers' surplus.

Given the demand and supply functions as $D(q)$ and $S(q)$ respectively, the suppliers' surplus is given by $\int_0^k (P(q) - S(q))d$ where k is the quantity at the equilibrium price.

Marginal cost

When a business entity is producing a certain number of units, there are costs associated with production of those units. If the entity increases the number of units it produces the cost of production will also increase since there might be need to employ extra employees, increase the raw materials and increase the units of electricity used in production. The extra cost is referred to as **marginal cost**. If the marginal cost function is $M(q)$ and a company increases the units of production from $q = r$ to $q = t$ the increase in total manufacturing cost will be $\int_s^t M(q)d$.

Example 1

A local authority estimates that q number of maize bags in thousands will be demanded when the price is $p = -0.2q^2 + 1$ dollars per maize bag. They also estimate that the q number of maize bags will be supplied when the price is $p = 0.3q^2 + q + 6$. Determine the supplier's surplus at the point of equilibrium.

Solution

The demand and supply functions are $D(q) = -0.2q^2 + 100$ and $S(q) = 0.3q^2 + q + 60$ respectively.

At equilibrium the supply is equal to the demand thus $S(q) = D(q)$.

$$0.3q^2 + q + 60 = -0.2q^2 + 100$$
$$0.5q^2 + q - 40 = 0$$

Solving this quadratic equation we get $q = 8$ and $q = -10$.

We take $q = 8$ since it is unreasonable to have -10 bags of maize.

These curves can be plotted as follows.

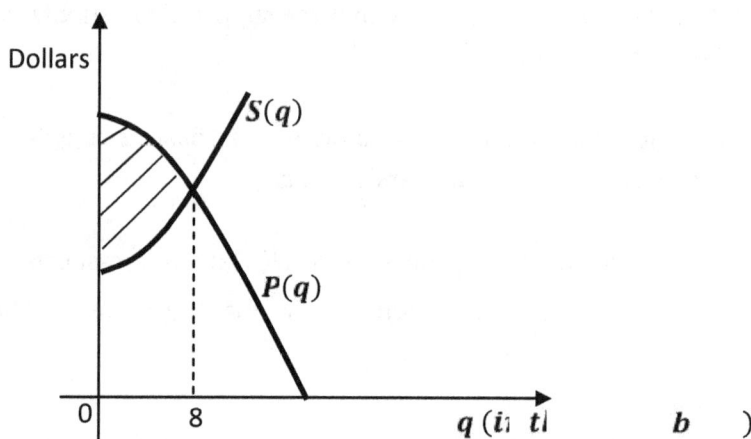

The shaded region represents suppliers' surplus.

We can find the area of the shaded region as follows.

$$\int_0^8 (P(q) - S(q))d$$

$$\int_0^8 (-0.2q^2 + 100) - (\mathbf{0.}3q^2 + q + 60)d$$

$$\int_0^8 (-\mathbf{0.}5q^2 - q + 40)d = \left| -\frac{0.5}{3}q^3 - \frac{q^2}{2} + 40q \right|_0^8$$

$$= \left(-\frac{0.5}{3}(8)^3 - \frac{8^2}{2} + 40(8) \right) - \left(\frac{0.5}{3}(0)^3 - \frac{(0)^2}{2} + 40(0) \right)$$

$$= 202.70 - 0 = 2 \quad .7 \quad \text{Dollars}$$

Example 2

The ministry of trade estimates the demand function for a certain type of computers to be $D(q) = 3(-q^2 + 2\ )$ Dollars per computer. Find the total amount of money that the consumers are willing to spend for five computers.

Solution

The total money that the consumers are willing to spend for y units is the area under the curve from when $q = 0\, u$ to when $q = y\, u$. this is equivalent to $\int_0^y D(q)d$

In this case the total money is $\int_0^5 3(-q^2 + 270)d$

$$= 3\left[\frac{-q^3}{3} + 270q\right]_0^5$$

$$= 3(-\frac{5^3}{3} + 270(5) - 3(-\frac{0^3}{3} + 270(0))$$

$$= 3925 \text{ Dollars}$$

Example 3

A company manufactures 10 tanks of acid in a day. Its marginal cost function is $4q^2 - 4\ + 2$. This company is considering increasing the number of tanks it manufactures to 15. Calculate the increase in total cost of production.

Solution

The increase in total cost of production is

$$\int_1^1 (4q^2 - 40q + 200)d$$

$$= \left[\frac{4q^3}{3} - 20q^2 + 200q\right]_1^1$$

$$= \left(\frac{4(15)^3}{3} - 20(15)^2 + 200(15)\right) - \left(\frac{4(10)^3}{3} - 20(10)^2 + 200(10)\right)$$

$$= (4500 - 4500 + 3000) - (1333.30)$$

$$= 1666.70 \text{ Dollars}$$

Example 4

Michael is considering two investment plans. The first plan generates money at a rate of $P(t) = 8 + t^2$ while the second plan generates money at a rate of $R(t) = 1 + t$ where t is time in years from now. After how money year will the first plan exceed the second plan?

If Michael chooses the first plan, calculate the extra money that he could have accumulated before the first plan exceeds the second plan had he chose the second plan.

Solution

The first investment plan exceeds the second investment plan at a point where $P(t) = R(t)$.

$$80 + t^2 = 100 + t$$
$$t^2 - t - 20 = 0$$

Solving this quadratic equation we get $t = -4$ and $t = 5$

Since we cannot have -4 years, the first investment plan exceeds the second investment plan after 5 years.

Had he chose the second plan, he could have accumulated extra $\int_0^5 (R(t) - P(t))d$ dollars before five years.

$$\int_0^5 (R(t) - P(t))d = \int_0^5 (100 + t - (80 + t^2))d$$

$$= \int_0^5 (20 + t - t^2))d$$

$$= \left[20t + \frac{t^2}{2} - \frac{t^3}{3} \right]_0^5$$

$$= \left(20(5) + \frac{(5)^2}{2} - \frac{5^3}{3} \right) - \left(20(0) + \frac{0^2}{2} - \frac{0^3}{3} \right)$$

$$= 70.80 - 0 = 70.80 \text{ Dollars}$$

Calculus 104.29 Appendix items, Basic Set Theory, Algebraic Rules

Consider a University having students in first year, second year and third year. These students can be grouped according to their of year study, their gender or according to their courses of study. Whichever way they are grouped, each category is referred to as a set. **A set** is collection of distinct items called the elements of a set, having a certain characteristic. Let's say that we group students according to their year of study and according to their course of study. We will have a set of students taking a degree in mathematics and a set students in the first year among the distinct sets.

We have students the in first year taking in mathematics too. These are the elements shared by the two sets. The set of shared elements is called the intersection of the sets. The intersection of set A and set B is denoted by $A \cap B$.

Let's define some of the terms used in set theory.

Subset
If A and B are two sets and all elements of B are also elements of A, then B is a subset of A.

Power set
This is the set of all subsets of a set.

Union of a sets ($\cup$)
If A and B are two sets the union of A and B is the set containing the elements of A and the elements of D.

Intersection of sets ($\cap$)
This is the set having elements common to all the sets being considered

Universal set
This is the set having all elements one can thing off.

Complement of a set
The complement of a set A (denoted A^C) is the set of elements that are in the universal set but on in A.

Algebraic rules

There are rules that govern addition, subtraction and multiplication of real numbers, variables and algebraic expressions.

In this lesson we will list these rules.

We will let j, k and l be real numbers, variables or algebraic expressions.

1. Commutative rule under addiction and subtraction.
$$j \pm k = k \pm j$$

2. Commutative rule under multiplication
$$j \times k = k \times j$$

3. Associative rule under addition
$$(j + k) + l = j + (k + l)$$

4. Associative rule under multiplication
$$(j \times k) \times l = j \times (k \times l)$$
Note that associative rule is not under subtraction since, $(j - k) - l \neq j - (k - l)$

5. Distributive rule of addition over multiplication
$$(j + k) \times l = (j \times l) + (k \times l)$$

6. Additive identity
Additive Identity of a real number j is the number that gives j after addition.
The additive identity is 0 since $j + 0 = j$.

7. The multiplicative identity is 1
$$j \times 1 = j$$

8. The additive inverse of a number **j** is **$-j$**.
$$j + (-j) = 0$$

9. Equality in addition, subtraction, multiplication and division
If $j = i$, and k is a real number, then $j \pm k = i \pm k$
$$jl = il$$

$$\frac{j}{k} = \frac{i}{k} \ s \ l\!\iota \quad a \quad k \neq 0$$

10. Equality in addition

If $j = i$, and k is a real number, then $j + k = i + k$

11. Symmetry of real numbers

$$j = j$$

12. Reflexivity in real numbers

If $j = k$ then $k = j$

13. Zero, a multiplicative anhillator

$$j.0 = 0.j = 0$$

Example 1

$A = \{j, k, l\}$. Find the power set of A.

Solution

The power set of set A is the set of all subsets of A. These subsets includes the empty set and the A itself.

$$P(A) = \{\{j, k, l\}, \{j, k\}, \{k, l\}, \{j, l\}\{j\}, \{k\}, \{l\}, \{\emptyset\}\}$$

Example 2

A college has students in the first year and the second year. The total number of students in the first year is 320. The students in second year are 240. 140 students in first year take degree in mathematics while 75 students in the same year take a degree in statistics. Thirty-two students in the second year take a degree in Actuarial Science while 120 students in the same year take a degree in Statistics. The college offers three courses in each year of study; a degree in mathematics, a degree in Statistics and a degree in Actuarial Science. In the second year 90% of the students passed their final exam. Twenty five percentof students who failed in second year take a degree in mathematics. Ninety five percent of students in the second year taking a degree in statistics passed their exam.

a) How many students in the first year do not take either a degree in mathematics orin statistics?

b) How many students in the second year taking a degree in Actuarial science passed their final exam?

Solution
The students in the entire college is the universal set.

Let the number of students in first year be $|A|$ and the number of students in second year be $|B|$ then $|A| = 320$ and $|B| = 240$.

Let students in the first year taking degree in Mathematics be the set C.

Let students in the first year taking degree in Statistics be the set D.

Let students in the first year taking degree in Actuarial science be the set E.
C, D and E are subsets of A.

Let students in the second year taking degree in mathematics be the set F.

Let students in the second year taking degree in Statistics the set G.

Let students in the second year taking degree in Actuarial science be the set H.

The students in the first year who do not take either degree in mathematics or degree in Actuarial science is the set $(C \cup E)^C$

Since C and E are subsets of A, $|(C \cup E)^C| = |A| - |C| - |E| - |C \cap E|$
$$= 320 - 140 - 75 - 0$$
$$= 105 \text{ Students}$$

b) Students in the second year is the set B. $|B| = 240$

Students in the second year taking degree in Actuarial science is the set H.

H is a subset of B.
Let the students in second year who fail exam be the set Q.
Let the students in second year who passed exam be the set S.
Let the students in second year who fail exam and taking actuarial science be the set R.
Let the students in second year who pass exam and taking actuarial science be the set Y.
Let the students in second year who pass exam and taking actuarial science be the set U
Let the students in second year who fail exam and taking degree in mathematics be the set T.
Let the students in second year who pass exam and taking degree in statistics be the set V.
Let the students in second year who fail exam and taking degree in statistics be the set W.

$$|S| = 90\% \text{ of } 240 = 216 \text{ Students}$$

$$|Q| = 10\% \text{ of } 240 = 24$$

$$|T| = 25\% \text{ of } 24 = 6 \text{ Students}$$

$$|W| = 5\% \text{ of } 120 = 6 \text{ Students}$$

W, R and T are subsets of Q, $|R| = |Q| - |W| - |T| - |W \cap T|$

$$= 24 - 6 - 6 - 0 = 12 \text{ Students}$$

We want to find $|Y|$

Y and R are subsets of G.

$$|G| = 32$$

$|Y| = |G| - |R| - |Y \cap R|$ Since a student can fail and at the same time pass an

Exam, $|Y \cap R| = 0$.

$$|Y| = 32 - 12 - 0 = 20 \text{ Students}.$$

Example 3

Simplify $3x(x^2 + 2x)$

Solution

Since distributive rule of addition over multiplication hold for expressions,

$$3x(x^2 + 2x) = (3x \times x^2) + (3x \times 2x)$$
$$= 3x^3 + 6x^2$$

Example 4

Find the additive inverse of $2x - 3x^2$

Let $2x - 3x^2 = k$ then its inverse is $-k$

$= -(2x - 3x^2) = -2x + 3x^2$

Section 104 Conclusion

In this lesson, we have discussed some methods of finding definite integrals. We then moved further to look at the applications of these definite integrals in business world and economics. We have finalized by looking at set theory and algebraic rules.

Section 104: Glossary

- **Annuity**: It is discrete amount of money deposited to an account at equal intervals of time for a given period of time earning a certain interest within the period.
- **Antiderivative or indefinite integral**: is a term used to imply the reverse process of determining a derivative.
- **Arithmetic series**: A series where any consecutive terms have a common difference
- **Asymptotes**: These are straight lines where the graph approaches but does not intersect nor touch
- **Circle**: A circle is a collection of points that are equidistant from a fixed point called the center of the circle.
- **Continuity**: A function $y = f(x)$ is said to be continuous at a point say, $x = a$ if

    (i). The function is defined at that point. That is $f(a)$ exists

    (ii). $\lim_{x \to a} f(x) = f(a)$
- **Continuity on an interval**: A function is continuous on an interval if it is continuous at all points within the interval
- **Continuity on real number set**: A function is continuous on the real number set if it is continuous at all points within the real number set
- **Cost function**: Is a function that determines the cost of a commodity that has both fixed and variable function
- **Demand**: It is the quantity of the commodity that consumers are willing and able to buy at certain price.
- **Depreciation**: This is a reduction of value of an item
- **Derivative of a function**: Derivative of a function $f(x)$ is the limit

    $$\frac{\lim_{h \to 0} \square \left(f(x+h) - f(x) \right)}{h} \quad \text{if it exists.}$$
- **Differentiability**: A function is differentiable at a point if the derivative exists at that point
- **Differentiation**: It is the process of determining the derivative of the curve with respect to a given variable.
- **Ellipse**: Is a collection of points where the sum of the distances from each point to two fixed points called the focus is the same for all points.
- **Exponential function**: This is function of the form $f(x) = ab^{kx}$ where a, b and k are non-zero contents with $b \neq 1.$
- **Future value of annuity**: It is the total money deposited plus the interest that has accumulated over a certain period when a person invests in annuities.
- **Geometric series**: A series where any consecutive terms have a common ratio
- **Higher derivatives**: These functions that results from differentiating a function more than once
- **Hyperbola**: A hyperbola is a set of points whose positive difference between the distances from the one point to the fixed points called the foci is the same

- **Intercepts**: These are points on the main axes where the graph passes
- **Limit**: The limit of a function $f(x)$ as x approaches a number a is a number L such that the values of $f(x)$ approaches L when values of x are taken closer and closer to a
- **Linear function**: Is a function that is in the form $ax + by + c = 0$
- **Logarithmic function**: This is function of the form, $f(x) = \log_a x$ w here a is a non-negative constant not equal to 0 nor 1.
- **Maclaurin series**: It is a Taylor series evaluated at $x = 0$.
- **Marginal cost**: It is the additional cost that an entity incurs when to increases the number of units it produces.
- **Optimization**: It refers to maximization and minimization of functions or variables
- **Parabola**: A parabola is a collection of points is equidistant from both the central point, called the focus, as the line called the directory
- **Power set**: It is the set of all subsets of a certain set
- **Quadratic function**: This is a function which is given by, $f(x) = ax^2 + bx + c$ where, a, b, c are constants with a being non zero
- **Rational function**: This is a function in the form $g(x) = \dfrac{f(x)}{h(x)}$ where $h(x)$ can never be zero and $f(x)$ and $h(x)$ are polynomials
- **Rules of differentiation**: Different ways of determining derivatives
- **Sequence**: It is a pattern of numbers, letters, or items. Here, we are mostly interested in sequence of numbers.
- **Series**: It is a sum of terms in a sequence.
- **Set**: It is collection of distinct items called the elements of a set, having a certain characteristic.
- **Slope**: This is the rise divided by the run of the graph
- **Square root function**: This is a function where the variable is under a square root function
- **Supply**: It is the quantity of the commodity that the suppliers can supply at a certain price.
- **Taylor's Polynomial**: Taylor polynomial of degree n is given by

$$P_n(x) = f(a) + (x-a)f'(a) + \frac{(x-a)^2}{2!}f''(a) + \frac{(x-a)^3}{3!}f''(a) + \cdots$$
$$+ \frac{(x-a)^n}{n!}f^n(a)$$

www.ingramcontent.com/pod-product-compliance
Lightning Source LLC
Chambersburg PA
CBHW051219200326
41519CB00025B/7173

* 9 7 8 1 9 4 4 3 4 6 6 2 1 *